BIOLOGICAL WASTEWATER TREATMENT AND RESOURCE RECOVERY

Edited by **Robina Farooq** and **Zaki Ahmad**

Biological Wastewater Treatment and Resource Recovery

http://dx.doi.org/10.5772/62795
Edited by Robina Farooq and Zaki Ahmad

Contributors

Asmita Gupta, Indu Shekhar Thakur, Adey Feleke Desta, Joyce Nzioki, Solomon Mania, Francesca Stomeo, Ajay Bissessur, Terelle Ramcharans, María Noel Cabrera, Ahmed Al-Busaidi, Mushtaque Ahmed, Esra Sündüz Yiğittekin, Sadık Dincer, Dragan Milicevic, Slavisa Trajkovic, Milan Gocic, Periyasamy Sivagurunathan, Gopalakrishnan Kumar, Arivalagan Pugazhendhi, Guangyin Zhen, Kobayashi Takuro, Kai-Qin Xu, Abdurahman H Nour, N.H. Azhari, Tonci Rezic, Iva Rezic, Michaela Zeiner, Carla Escapa Santos, Ricardo Nuno de Coimbra, Sergio Paniagua Bermejo, Ana Isabel García Pérez, Marta Otero Cabero, Edson Luiz Silva, Eduardo Lucena Cavalcante de Amorim, Leandro Takano Takano Sader, Lucas Rodrigues Ramos, Shaon Ray Chaudhuri, Madhusmita Mishra, Sonakshi De, Biswajit Samal, Amrita Saha, Srimoyee Banerjee, Abhinandan Chakraborty, Antara Chakraborty, Sonali Pardhiya, Deepak Gola, Joyeeta Chakraborty, Sourav Ghosh, Kamlesh Jangid, Indranil Mukherjee, Mathummal Sudarshan, Rajib Nath, Ashoke Ranjan Thakur

CBS, Edition **2018**

Published by InTech

Janeza Trdine 9, 51000 Rijeka, Croatia

Publishing Process Manager Martina Usljebrka

Technical Editor SPi Global

Cover Designer InTech Design Team

Additional hard copies can be obtained from orders@intechopen.com

Biological Wastewater Treatment and Resource Recovery, Edited by Robina Farooq and Zaki Ahmad
p. cm.
Print ISBN 978-953-51-3045-1
Online ISBN 978-953-51-3046-8

Contents

Preface

Industrialization and rapid increase in global population resulted in the increase of hazardous chemicals including radioactive materials, pesticides, fertilizers, pharmaceuticals, petrochemicals, dyes, paints, heavy metals, surfactants and detergents in the environment. These organic and inorganic substances find their way via air or soil to water and are the constant source of toxicity to flora and fauna resulting in the extinction of species. One of the major challenges facing mankind today is the provision of clean water and food to a vast majority of the population around the world. Nature has an amazing ability to cope with small amounts of water pollution, but the treatment of billions of gallons of wastewater and sewage produced every day is required to be treated before releasing it back to the environment. Wastewater contains precious chemicals and substances, which can be recovered using different chemical and biochemical reactions. The main focus of the book is to describe effective methods for wastewater management and its treatment, reuse and recycling along with the recovery of valuable substances and energy.

Volume I focuses on the bioremediation of wastewater and is divided into four sections. Section 1 'Wastewater, Management and Monitoring' emphasizes on the micropollutants entering into the environment after conventional wastewater treatment facilities of industrial, agricultural and domestic wastewaters. The occurrence of these persistent pollutants poses deleterious effects on human and environmental health. The fate and persistence of these chemicals after conventional treatment processes are discussed. Simple solution for the treatment of wastewater and recovery of water as resource using microbiological method is a viable option. This increases biomass and reduces water, air and soil pollution. Mitigating environmental risks of wastewater reuse for agriculture is also discussed with the support of experimental studies. An interesting book chapter on 'Micro-Based Strategy for Plant Nutrient Management' is included in this section, which highlights the importance of slow leaching of nutrients for plant uptake. It provides modest solution to readers and farmers for the use of mixture of microbial consortia in soil, which not only helps to reduce leaching of nutrients in groundwater but also reduces ground water pollution along with the cost of fertilizers.

Section 2 'Hazards and Treatment of Organic Compounds in Wastewater' describes studies about the hazards and treatments of wastewater containing antibiotics, pharmaceuticals, cyclic aromatic compounds and bleaching agents.

Persistent antibiotics in wastewater develop resistant microorganisms. This poses great financial and research burden to find alternate antibiotics. Their fate and treatment provides insight about the process. Similarly, the presence of pharmaceutical compounds and the emerging contaminants in wastewater cause physiological responses in nontarget organisms. This section covers the identification of efficient microorganisms, which is the key factor for biological treatment of wastewater containing such contaminants. Identification of relationship between *Bacteroidetes* and *Proteobacteria* for treatment of xenobiotic compounds in the wastewater of tanning industry conducted on field-scale reactor is an important part of this section. A comprehensive review of pulping technologies and biological treatment of wastewater is included. Here, we also focus on known and emerging pollutants.

Section 3 'Biofilms for Wastewater Treatment' is designated for the application of microbial culture for bioabsorption of metals and treatment of surfactants. Parameters effecting biosorption of metals on biofilm are optimized using pilot-scale horizontal tubular bioreactor. The study about electrocoagulation and biological treatment of laundry wastewater investigates the possibility of recycling and reusing wastewater from laundry run-offs.

Section 4 'Bioenergy as Resource Recovery' draws the attention of readers and researchers towards the recovery of biogas, hydrogen, volatile fatty acids and alcohols during anaerobic treatment of wastewaters from carbohydrates present in wastewaters. This covers the discussions about the synthesis of biofuel using anaerobic fluidized bed reactor for hydrogen synthesis from glucose-based wastewater. Another interesting study is the production of biogas and its performance evaluation using ultrasonic membrane anaerobic system (UMAS). Integrated technology of UMAS is an attractive solution for treatment of palm oil wastewater along with resource generation. This section also describes two important pathways for the production of hydrogen, which are light and dark fermentation reactions. Recent advancements on biohydrogen production technologies from wastewater with respect to inoculum development, process optimization, scale-up and challenges are discussed in detail. Bioremediation of wastewater is the low-cost solution. However, its efficiency is effected by environmental conditions. Therefore, physico-chemical treatment of wastewater is another efficient option, which will be covered in Volume II.

I would like to express my gratitude to Prof. Dr. Zaki Ahmad who started to work with me as coeditor. Prof. Zaki passed away during his work on this book. His efforts and contributions are highly appreciated and his services as book editior are highly acknowledged.

I would like to thank Ms. Martina Usljebrka, Publishing Process Manager, for enabling me to publish this book. I want to thank my husband Prof. Dr. Saleem Farooq Shaukat, my daughter Kinza Farooq, sons Abdul Basit and Faisal Farooq and grandchildren Zoha Fatima and Aarib Basit who kept me motivated to accomplish this work. I am grateful for my father Mr. Muhammad Mukhtar and my mother Mrs. Rafia Mukhtar, my sisters and my brothers who always supported and encouraged me throughout my life.

Prof. Dr. Robina Farooq
Department of Chemical Engineering
COMSATS Institute of Information Technology
Lahore, Pakistan

Prof. Dr. Zaki Ahmad
COMSATS Institute of Information Technology
Lahore, Pakistan

Note from the Publisher

It is with great sadness and regret that we inform the contributing authors and future readers of this book that editor Prof. Zaki Ahmad passed away during his work on publications and before having a chance to see them.

Prof. Ahmad was the InTech's long-term collaborator and edited his first book with us in 2011 *Recent Trends in Processing and Degradation of Aluminium Alloys,* followed by publications *Aluminium Alloys - New Trends in Fabrication and Applications, New Trends in Alloy Development, Characterization and Application* and *High Temperature Corrosion*. This fruitful collaboration continued until his final days when he was acting as a coeditor on the books *Biological Wastewater Treatment and Resource Recovery* and *Physico-Chemical Wastewater Treatment and Resource Recovery.*

We would like to acknowledge Dr. Zaki Ahmad's contribution to open access scientific publishing, which he made during his 6 years of dedicated work on edited volumes, and express our gratitude for his always pleasant cooperation with us.

InTech Book Department Team

Section 1

Wastewater Management and Monitoring

Chapter 1

Treatment of Organic Recalcitrant Contaminants in Wastewater

Asmita Gupta and Indu Shekhar Thakur

Additional information is available at the end of the chapter

http://dx.doi.org/10.5772/66346

Abstract

Research has shown that a myriad of contaminants enter the environment through industrial and domestic sources on a daily basis. The biodegradable compounds often get degraded or mineralized by various physical, chemical or biological processes, whereas the recalcitrant organic contaminants either are transformed or get dispersed and persist in the receiving environments, and to an extent much greater than was earlier estimated. Many chemical compounds that were not previously included as pollutants can now be detected at much higher concentrations globally. The effect of most of these emerging contaminants on human and environment health is still unknown. Therefore, there is an urgent need to study the fate of these persistent compounds so as to better understand and manage their ecological and health effects.

Keywords: Wastewater, organic contaminants, recalcitrant, biodegradation, sorption

1. Introduction

Water adversely affected in quality by anthropogenic activities is, typically called wastewater. Wastewater is generally collected and treated by various processes at centralized facilities, referred to as wastewater treatment plants (WWTPs). There can be several sources contributing towards wastewater generation, including domestic, industrial and agricultural. As there are various sources of wastewater generation, so are the compounds present in them. Wastewater, thus, is a cocktail of chemicals—the class, structure, biodegradability, toxicity and human and environmental impact of most of which are still unknown.

Some of the wastewater contaminants, including aromatics, pharmaceuticals, pesticides,33 chlorinated congeners and plasticizers, pose deleterious effects on human and environmental health, even at trace levels [1]. Some of their harmful effects include impairment and/or abnormality in physiological processes, including reproductive impairment, increased risk of cancer in aquatic and terrestrial species, development of antibiotic-resistant bacterial strains and increase in effluent toxicity post-treatment plausibly owing to the synergistic or antagonistic toxic effects of such recalcitrant chemical mixtures. Still unknown are the environmental effects of many emerging contaminants.

While most of the easily degradable wastewater contaminants are removed by conventional treatment methods, compounds that remain even in the treated effluent are recalcitrant and hence persist in the receiving environments, causing environmental and health problems. Low concentrations of such recalcitrants in large volumes of wastewater make their efficient treatment and removal very difficult by the conventional treatment processes including activated carbon, chemical precipitation, ionic exchange resins and membrane filtration [2]. Such processes have other disadvantages such as high plant operation and maintenance cost, accumulation and disposal issues of concentrated sludge, use of excessive chemicals, low sensitivity towards target compounds and accumulation of concentrated sludge and their disposal problems [3]. Removal of some of the organic recalcitrants is not effected even by the traditional biological processes, including activated sludge and trickling filters, employing microorganisms as these biorecalcitrants may result in death of the microbial population, thus reducing the efficiency of or halting the treatment process. Advanced treatment methods such as a pre-separation step or post-treatment of recalcitrants using potent and specialized microbial strains need to be employed for the efficient removal of such persistent organic pollutants from effluent [2].

Hence, there is a need for better understanding of the occurrence, behaviour and fate of organic contaminants during sewage treatment processes. The present paper reviews literature about the fate of some of the recalcitrant organic contaminants during the various treatment processes.

2. Status of wastewater generation

Better management of wastewater at regional and global level requires up-to-date information on the status of sewage generation and treatment. Globally, a complete sewage generation and treatment data are available for only 55 countries, 37% of it being recent (2008–2012) [4]. There is a generation of about 15, 644 millions litre per day (MLD) of sewage from 35 metropolitan cities in India, out of which only 8040 MLD (51.4%) is the existing treatment capacity. While 3800 MLD is the municipal sewage generation in the national capital region of Delhi, the city has a treatment capacity of only about 2300 MLD. Rest 31% sewage is discharged into the environment untreated [5].

3. Wastewater treatment processes

Various processes are employed for the removal of wastewater contaminants depending on their type and level in the influent. Municipal wastewater is mostly treated in sewage treatment plants (STPs) which use various treatment processes including physical, chemical and biological. Wastewater treatment and discharge are done according to regional and national regulations and standards. Wastewater treatment is done with the purpose of producing a pollutant- and toxicity-free effluent which can safely be discharged into the environment [6]. Three main stages are involved in wastewater treatment, viz., primary or physic-chemical, secondary or biological and tertiary or advanced treatment.

a. *Primary treatment* involves physical separation of heavy solid particles gravimetrically and oil and other lighter floating materials mechanically in settling basins called primary-settling tanks or primary clarifiers. The remaining liquid wastewater is pumped to the next treatment tank for secondary treatment.

b. *Secondary treatment* involves the removal of dissolved and suspended biological component by means of an indigenous microbial population, which is removed prior to release of the treated water into the environment or tertiary treatment stage. It is carried out in secondary treatment chambers such as aeration tanks or bioreactors. In the presence of sufficient oxygen supplied through aeration pumps, the indigenous microflora degrades the soluble organic fractions while segregating the less soluble components into flocs. Secondary treatment may include either *fixed-film* or **attached growth** systems such as trickling filters, rotating biological contactors and bio-towers, where the sewage passes over the surface of attached biomass, or *suspended-growth* systems including activated-sludge process, where sewage is mixed with microbial biomass. While the latter type of secondary treatment system has a lower space requirement for wastewater treatment, requires less space for treatment, the fixed-film systems are better able to acclimatize to sudden microbial changes and have a higher removal rate of organic matter and suspended solids [7–9].

c. *Tertiary treatment* includes any advanced wastewater treatment methods beyond the primary and secondary treatment, before discharge of wastewater in the receiving environment.

The most important aerobic treatment system is the activated-sludge process, based on the maintenance and recirculation of a complex biomass composed by micro-organisms able to absorb and adsorb the organic matter carried in the wastewater. Other biological treatment processes such as expanded granular sludge bed (EGSB) reactor and upflow anaerobic sludge blanket (UASB) are also employed for wastewater treatment. Synthetic membranes and micro-filtration are now commonly being used as tertiary treatment technologies.

4. Fate of organic recalcitrant contaminants in wastewater treatment

4.1. Pathways of contaminant removal

There has been a radical increase in the occurrence and concentration of organic contaminants in wastewater and sludge as a result of an increase in the demand and industrial production of synthetic organic chemicals. Point discharge sources including discharges from industrial users or manufacturers and diffuse discharge sources such as commercial and domestic premises or run-off after aerial deposition are some of the major contributors to the loading of organic contaminant in sewage. The following are some of the pathways (**Figure 1**) through which organic contaminants may be transformed or degraded during sewage treatment:

- Air stripping
- Biodegradation
- Chemical degradation
- Sorption
- Volatilization

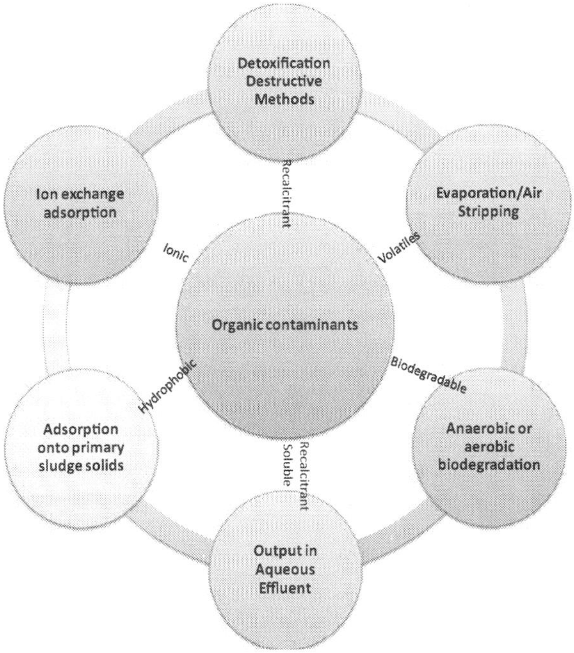

Figure 1. Some of the pathways involved in transformation of organic contaminant in wastewater treatment.

http://dx.doi.org/10.5772/66346

While some compounds may get completely degraded or mineralized in the process of treatment, some others are partially degraded and form breakdown products and a few other recalcitrant compounds may remain unaffected and persist in the effluent even after treatment. The occurrence of these synthetic organic contaminants in wastewater may be either in solution or sorbed onto solids. The hydrophobic or lipophilic nature of many organic contaminants result into their getting adsorbed on solid particles during wastewater treatment, eventually resulting in their accumulation in the sludge solids, sometimes at concentrations much higher than in the untreated wastewater [10, 11].

Structural composition of the organic residues may also provide information about their biodegradation pathways. For instance, biodegradation of unbranched and long-chained hydrocarbons is easier as compared to the short-chained or highly branched molecules. Biodegradation of unsaturated aliphatic compounds is generally more favoured than their saturated analogues. Molecules having highly polar groups and linkages tend to react by nucleophilic displacement (such as hydrolysis) [12]. Petrasek et al. [13] reported the association of recalcitrant and toxic chloro-organic pentachlorophenol (PCP) with the sludge solids, and considerable degradation of phenolic compounds having polar groups.

4.2. Processes involved in contaminant removal

Several researches have been made to study the removal efficiency of various contaminants by different wastewater treatment processes. Partitioning of hydrophobic contaminants of influent onto settled primary sludge solids may take place during the primary sedimentation process in the primary clarifiers. Bulk organic components of wastewater such as cellulose, proteins and carbohydrates get biodegraded during the secondary treatment involving aerobic processes such as trickling filters, activated-sludge process, oxidation ponds or anaerobic processes resulting in sludge digestion. Transformation or loss of some of the synthetic recalcitrant organic contaminants may also take place during the secondary treatment processes. Polysaccharides, proteins and fats occur in two phases during the anaerobic digestion process. First phase (acid phase) involves hydrolysis of polysaccharides to form mono- and disaccharides, of proteins to form amino acids, and of fats resulting in the formation of long-chain fatty acids, and volatile acids such as formic, acetic and butyric acid. Second phase (methanogenic phase) results in the reduction of the volatile acids to methane and carbon dioxide [12, 14]. In one study involving a generalized model for the presentation of fate of organic compounds in an activated-sludge process, it was demonstrated that the phase distribution of xenobiotic chemicals depended quantitatively upon their physico-chemical properties and the operating conditions of wastewater treatment. The study also showed the removal of hydrophobic chemicals of wastewater, mostly by the process of sorption onto sludge particles followed by their transfer to the sludge-processing units. Meanwhile, advective transport into the final effluent and biodegradation was shown to be the common mechanism for the removal of hydrophilic compounds of wastewater. The model also predicted an increase in the effluent concentration of complex organics such as substituted phthalates, high molecular weight (HMW) polycyclic aromatic hydrocarbons (PAHs) and dioxins with increasing solids retention time (SRT) during the operation of wastewater treatment plant [15].

4.3. Common classes of contaminants found in wastewaters

Although wastewaters contain a multitude of contaminants, yet they can be broadly grouped under different classes on the basis of their chemical structure. A total of 129 specific pollutants including heavy metals and specific organic chemicals have been defined by the US Clean Water Act as "Priority Pollutants". Municipal Environmental Research Laboratory (MERL), US EPA, conducted a comprehensive research programmes on the occurrence and fate of priority pollutants present in wastewater and sludge. The study assessed the fate and behaviour of 22 harmful organics including phenols, pesticides, poly aromatic hydrocarbons and phthalates in the conventional water treatment systems and demonstrated up to 95–98% removal of organic compounds from the liquid phase. Many such organic compounds were found to have been partitioned onto the solid phases of primary and return activated sludges. Similar results were reported in other studies as well [16, 17]. In one study, the highest degree of enrichment of PAHs was observed in the primary sludge and phthalates such as bis-(ethylhexyl) and di-n-octyl phthalate were found to be among the most recalcitrant compounds present in wastewaters [13]. Wild and Jones [18] reported the occurrence of volatile chemicals, such as benzene, in sewage sludge, possibly as a result of their sorption over organic substances present in the sludge. Based on the reported literature, the following description discusses the fate of some common classes of organic compounds occurring in wastewaters (**Figure 2**).

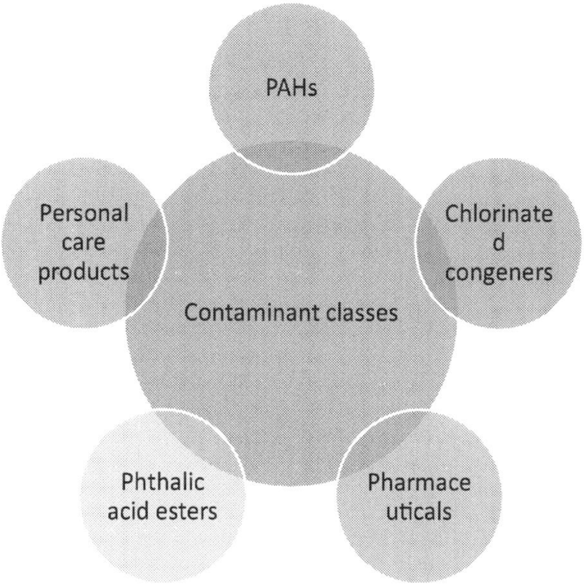

Figure 2. Classes of organic contaminants commonly found in wastewater.

4.3.1. *Phthalic acid esters*

Phthalates have a high environmental significance owing to their high production volumes as well as their eco-toxicological effects especially on aquatic fauna including molluscs, crustaceans and amphibians. They have been reported to cause biological effects even at very low levels of exposure, varying in the range of ng L^{-1} to µg L^{-1} [19, 20]. Microbial degradation of phthalates under aerobic and anaerobic conditions has been previously reported [21]. The difference in the biodegradability of various phthalates could possibly be due to the steric effect of their side ester chains that hinders the binding of hydrolytic enzymes to the phthalates thus inhibiting their hydrolysis [22]. In a previous study on the occurrence of phthalates in raw and treated wastewater of WWTPs, it was found that most of the studied phthalates were present in post-treated water samples, bis(2-ethylbenzyl) phthalate (DEHP) being the most abundant. Also, biotransformation and adsorption onto sludge solids (that directly depend on the molecular weight and lipophilic nature of the compound) were shown to be the possible pathways of phthalate removal from liquid phase during wastewater treatment [23]. Roslev et al. [24] studied the degradation of four different phthalic acid esters in an activated-sludge process, and showed an almost 96% association of DEHP (showing the least biodegradation among the four phthalates) with the wastewater suspended solids. The study also revealed a 7–9% recovery of the influent phthalate esters in the effluent. Also, aerobic and anoxic-denitrifying conditions were found to be less favourable for biodegradation of phthalate esters as compared to the alternating aerobic-anoxic conditions.

4.3.2. *Polycyclic aromatic hydrocarbons*

PAHs are among the most mutagenic, carcinogenic and toxic class of organic contaminants some of which have also been included in the US-EPA and EU list of priority pollutants [25]. The presence of PAHs in the environment is commonly attributed to various anthropogenic activities such as petroleum refining, power and heat generation from coal production, and chemical manufacturing [26]. A study on the fate of PAHs and other volatile organic compounds (VOCs) during wastewater treatment by the conventional activated-sludge process (CASP) and the membrane bioreactors (MBRs) concluded that aromatic VOCs were removed mainly by volatilization and with comparable removal efficiencies for both treatment processes, that is, CASP and MBRs. On the other hand, removal efficiency for PAHs was found to be enhanced in case of MBRs [27]. In another study conducted by Zhang et al. [28], the occurrence, behaviour and fate of 18 PAHs in a coking wastewater treatment plant was investigated and it was found that mostly high molecular weight PAHs were present in the raw coking wastewater, while 3–6 ring PAHs were the predominant PAHs detected in the effluent. There was detection of PAHs such as pyrene, phenanthrene and fluoranthene in the gas samples and pyrene, fluoranthene, chrysene and benzo[k]fluoranthene in sludge. While there was almost 97% removal for all the PAHs during treatment, the percent removal of PAHs from the liquid phase varied in a range of 47–92% in the biological stage. It was also observed that low molecular weight (LMW) PAHs were mostly removed in the aerobic tanks and following the mechanism of transformation, whereas their HMW counterparts were mainly removed in anaerobic tank. While transformation was observed to be the most common mechanism of removal of LMW PAHs from wastewaters, adsorption onto sludge solids was mainly responsible for the removal of HMW PAHs from the liquid phase.

4.3.3. Chlorinated congeners

Chlorinated congeners including polychlorinated biphenyls and polychlorinated pesticides are very toxic to human and environment health and are mostly added into the environment by industrial and domestic sources. Their presence has commonly been reported in wastewater, surface water bodies as well as in sediments. Biologically mediated reductive dehalogenation process is one of the common pathways of degradation of these chlorinated contaminants during wastewater treatment. The less investigated reductive dechlorination process has also been identified as one of the possible pathways for the transformation of specific contaminants during anaerobic digestion of sludge. Previous studies have reported the formation of intermediates such as 1, 2 ,4-trichlorobenzene and pentachlorobenzene, 1, 2, 4, 5-tetrachlorobenzene and final products such as dichlorobenzene isomers and 1, 3 ,5-trichlorobenzene during the reductive dechlorination of hexachlorobenzene. The formation of 2, 4-dichlorophenol and 4-chlorophenol as intermediates and phenol as the end product during reductive dechlorination of 2, 4-dichlorophenoxy acetate has similarly been reported [29].

While some of the chlorinated congeners such as polychlorinated biphenyls, have been known for long [30], some others have recently been documented as toxic contaminants including pharmaceuticals such as diclofenac and pesticides 4-hydroxychlorothalonil and clomazone [31, 32]. The detection of such chlorinated contaminants, some of which are also endocrine-disrupting and toxic to biota, in effluent and receiving water bodies is a matter of concern [33]. The concentrations of chlorinated congeners in effluent have been reported to be much lower than in the influent, indicating their efficient removal by various physical, chemical or biological processes operational during the treatment of wastewater [34]. Nevertheless, there have been reports indicating the presence of chlorinated contaminants such as triclosan and triclocarban in effluent of STPs, and eventually in the downstream water bodies and sediments [35, 36], thus pointing towards a need for upgradation of treatment mechanisms for their efficient removal. In a study conducted on the efficiency of aerobic and anaerobic processes in organic contaminant removal during treatment processes, it was concluded that a sequential system using a combination of both oxidative and reductive processes was probably the most efficient for the removal of recalcitrant organics. Highly chlorinated and volatile organohalogen compounds were found to degrade appreciably only under anaerobic conditions, while being resistant to oxidative degradation under aerobic conditions [37].

4.3.4. Pharmaceutical compounds

Pharmaceutical compounds are another class of emerging contaminants that have gained growing concerns in the past two decades mostly because of their less known health and environmental effects and ever-increasing usage and unchecked release into the environment. Metabolic excretion post consumption and improper disposal techniques are the main sources of these compounds in the environment. In a study conducted to investigate the presence of some common pharmaceutical compounds and fluoroquinolones (one of the "priority pollutants" having potential hazardous effects on the aquatic life) in two wastewater treatment plants in Spain, frequent detection of pharmaceuticals such as analgesics, anti-inflammatories and lipid regulators in effluent and incomplete elimination of most of the fluoroquinolones posttreatment was observed. The results also demonstrated higher efficiency of membrane bioreactor technique in removing pharmaceutical compounds as compared to the activated-sludge process [38]. Similar findings have been reported by other workers as well [39, 40].

4.3.5. *Personal care products*

There has been a recent concern over the toxic and ecological impact of personal care products (PCPs). Although there have been several reports on the assessment of concentrations of these chemicals in the environment [41–43], less work has been done to know their fate in the environment. In one assessment of the efficiency of various treatment processes for the removal of pharmaceuticals and personal care products, it was concluded that membrane bioreactor and activated-sludge process with nitrogen treatment were the most efficient processes for the treatment of such compounds [44].

5. Conclusion

Wastewater treatment facilities such as wastewater treatment plants, or domestic septic systems, which have been operating on the conventional technologies, are often inefficient in treating such a cocktail of compounds ranging from simple to complex and recalcitrant organic compounds. Thus, these centralized facilities, discharging treated effluent, which may still be contaminated with household chemicals, pharmaceuticals and biogenic hormones, into the environment end up being a source of pollutants for the receiving water bodies. Also, the sewage sludge generated at the STPs, often having a high accumulation of recalcitrant and hydrophobic contaminants, acts as a sink of such contaminants in the treatment facilities but a major source of organic recalcitrants when directly used as manure.

Such unchecked disposal and use of sewage and sludge into the environment or their direct application for domestic or agriculture purposes could lead to exposure of toxic contaminants to biological systems, possibly resulting in adverse metabolic responses. Advanced treatment technologies such as membrane bioreactors and sequential system using a combination of both oxidative and reductive processes were found to be more effective in the removal of various organic recalcitrant compounds. Therefore, implementation of such treatment technologies and addition of tertiary treatment techniques to the conventional methods, for the removal of such persistent contaminants,have become quintessential.

Thus, the occurrence of persistence organic contaminants in the effluent and sludge posttreatment and ambiguity about their fate pose a serious environmental challenge. Therefore, much research is still needed to identify the source, behaviour and sink as well as their ecological and health effects.

Acknowledgements

This paper was supported by the research grants of the Jawaharlal Nehru University, New Delhi, India. The author (Gupta A) also thanks the University Grants Commission (UGC), New Delhi, Government of India, for providing Research Fellowship.

Author details

Asmita Gupta* and Indu Shekhar Thakur

*Address all correspondence to: asmitagupta5@gmail.com

School of Environmental Sciences, Jawaharlal Nehru University, New Delhi, India

References

[1] Lai HT, Hou JH, Su CI, Chen CL. Effects of chloramphenicol, florfenicol, and thiamphenicol on growth of algae *Chlorella pyrenoidosa, Isochrysis galbana,* and *Tetraselmis chui*. Ecotoxicol. Environ. Safe. 2009; 72:329–334.

[2] Padmavathy S, Sandhya S, Swaminathan K, Subrahmanyam YV, Chakrabarti T, Kaul SN. Microaerophilic-aerobic sequential batch reactor for treatment of azo dyes containing simulated wastewater. Process Biochem. 2005; 40(2):885–890.

[3] Balaji T, Matsunaga H. Adsorption characteristics of As(III) and As(V) with titanium dioxide loaded Amberlite XAD-7 resin. Anal. Sci. 2002; 18:1345–1349.

[4] Sato T, Qadir M, Yamamoto S, Endo T, Zahoor A. Global, regional, and country level need for data on wastewater generation, treatment, and use. Agricul. Water Manage. 2013; 130:1–13. ISSN 0378-3774

[5] CPCB. Inventorization of sewage treatment plants, Control of urban pollution Series: CUPS//2015. Central Pollution Control Board, India. 2015. Available from: http://cpcb.nic.in/upload/NewItems/NewItem_210_Inventorization_of_Sewage-Treatment_Plant.pdf

[6] Khopkar SM. Environmental Pollution Monitoring and Control. New Delhi: New Age International; 2004. p. 299. ISBN81-224-1507-5.

[7] EPA. Package Plants. Wastewater Technology Fact Sheet. Washington, DC. Document no. EPA. 2000; 832-F-00-016.

[8] EPA. Primer for Municipal Waste Water Treatment Systems. Washington, DC. Document no. EPA. 2004; 832-R-04-001.

[9] Randall CW, Sen D. Full-scale evaluation of an integrated fixed-film activated sludge (IFAS) process for enhanced nitrogen removal. Water Sci. Technol. Elsevier; 1996; 33(12): 155–162. doi: 10.1016/0273-1223(96)00469-6

[10] Katsoyiannis A, Zouboulis A, Samara C. Persistent organic pollutants (POPs) in the conventional activated sludge treatment process: model predictions against experimental values. Chemosphere. 2006; 65(9):1634–1641.

[11] Rogers HR. Sources, behaviour and fate of organic contaminants during sewage treatment and in sewage sludges. Sci. Total Environ. 1996; 185:3–26.

[12] Scow KM. Rate of biodegradation. In: Lyman WJ, Reehl WF and Rosenblatt DH (eds). Handbook of Chemical Property Estimation Methods. Environmental Behaviour of Organic Compounds. New York, NY: McGraw Hill; 1982.

[13] Petrasek AC, Kugelman IJ, Austern BM, Pressley TA, Winslow LA, Wise RH. Fate of toxic organic compounds in wastewater treatment plants. J. Water Poll. Contr. Fed. 1983; 55(10):1286–1296.

[14] Mason CF. Biology of Freshwater Pollution, Harlow: Longman; 1981.

[15] Byrns G. The fate of xenobiotic organic compounds in wastewater treatment plants. Water Res. 2001; 35(10):2523–2533.

[16] Gupta A, Thakur IS. Biodegradation of wastewater organic contaminants using Serratia sp. ISTVKR1 isolated from sewage sludge. Biochem. Eng. J. 2015; 102:115–124.

[17] Harrison EZ, Oakes SR, Hysell M, Hay A. Organic chemicals in sewage sludges. Sci. Total Environ. 2006; 367(2–3):481–497.

[18] Wild SR, Jones KC. Organic chemicals entering agricultural soils in sewage sludges: screening for their potential to transfer to crop plants and livestock. Sci. Total Environ. 1992; 119:85–119.

[19] Call DJ, Cox DA, Geiger DL, et al. An assessment of the toxicity of phthalate esters to freshwater benthos: 1. Aqueous exposures. Environ. Toxicol. Chem. 2001; 20:1798–1804.

[20] Oehlmann J, Schulte-Oehlmann U, Kloas W, et al. A critical analysis of the biological impacts of plasticizers on wildlife. Philos. Trans. R. Soc. B. 2009; 364:2047–2062.

[21] Staples CA, Peterson DR, Parkerton TF, Adams WJ. The environmental fate of phthalate esters: a literature review. Chemosphere. 1997; 35:667–749.

[22] Liang D-W, Zhang T, Fang HHP, He J. Phthalates biodegradation in the environment. Appl. Microbiol. Biotechnol. 2008; 80:183–198.

[23] Clara M, Windhofer G, Hartl W et al. Occurrence of phthalates in surface runoff, untreated and treated wastewater and fate during wastewater treatment. Chemosphere. 2010; 78(9):1078–1084.

[24] Roslev P, Vorkamp K, Aarup J, et al. Degradation of phthalate esters in an activated sludge wastewater treatment plant. Water Res. 2007; 41(5):969–976.

[25] Arfsten DP, Schaeffer DJ, Mulveny DC. The effects of near ultraviolet radiation on the toxic effects of polycyclic aromatic hydrocarbons in animals and plants: A review. Ecotox. Environ. Safe. 1996; 33:1–24.

[26] Suess MJ. The environmental load and cycle of polycyclic aromatic hydrocarbons. Sci. Total Environ. 1976; 6:239–250.

[27] Fatone F, Di Fabio S, Bolzonella D, Cecchi F. Fate of aromatic hydrocarbons in Italian municipal wastewater systems: an overview of wastewater treatment using conventional activated-sludge processes (CASP) and membrane bioreactors (MBRs). Water Res. 2011; 45(1):93–104.

[28] Zhang W, Wei C, Chai X et al. The behaviours and fate of polycyclic aromatic hydrocarbons (PAHs) in a coking wastewater treatment plant. Chemosphere. 2012; 88:174–182.

[29] Schwartzenbach RP. Groundwater contamination by volatile halogenated alkanes: abiotic formation of volatile sulphur compounds under anaerobic conditions. Environ. Sci. Technol. 1985; 19:322–327.

[30] Gutierrez AG, McIntyre AE, Perry R, Lester JN. Behaviour of persistent organochlorine micropollutants during primary sedimentation of waste water. Sci. Total Environ. 1984; 39:27–47.

[31] Hug C, Ulrich N, Schulze T, Brack W, Krauss M. Identification of novel micropollutants in wastewater by a combination of suspect and non-target screening. Environ. Pollut. 2014; 184:25–32.

[32] Nelson ED, Do H, Lewis RS, Carr SA. Diurnal variability of pharmaceutical, personal care product, estrogen and alkylphenol concentrations in effluent from a tertiary wastewater treatment facility. Environ. Sci. Technol. 2011; 45:1228–1234.

[33] Krzmarzick MJ, Novak PJ. Removal of chlorinated organic compounds during wastewater treatment: achievements and limits. Appl. Microbiol. Biotechnol. 2014; 98(14): 6233–6242.

[34] Langford KH, Lester JN. Fate and behavior of endocrine disrupters in wastewater treatment processes. In: Lester JN, Birkett JW (eds) Endocrine Disrupters in Wastewater and Sludge Treatment Processes. London: CRC Press; 2003. 103–143.

[35] Zhang Q-Q, Zhao J-L, Liu Y-S, Li B-G, Ying G-G. Multimedia modeling of the fate of triclosan and triclocarban in the Donjian River Basin, South China and comparison with field data. Environ Sci: Process Impacts. 2013; 15(11):2142–2152.

[36] Gautam P, Carsella JS, Kinney CA. Presence and transport of the antimicrobials ticlocarban and triclosan in a wastewater-dominated stream and freshwater environment. Water Res. 2014; 48(1):247–256.

[37] Zitomer DH, Speece RE. Sequential environments for enhanced biotransformation of aqueous contaminants. Environ. Sci. Technol. 1993; 27:226–243.

[38] Guedes-Alonso R, Afonso-Olivares C, Montesdeoca-Esponda S, Sosa-Ferrera Z and Santana-Rodríguez JJ. An Assessment of the Concentrations of Pharmaceutical Compounds in Wastewater Treatment Plants on the Island of Gran Canaria (Spain). Springer Plus. 2013; 2:24. http://www.springerplus.com/content/2/1/24

[39] Ö llers S, Singer HP, Fässler P, Müller SR. Simultaneous quantification of neutral and acidic pharmaceuticals and pesticides at the low-ng/l level in surface and waste water. J. Chromatogr. A. 2001; 911:225–234.

[40] Ternes TA. Occurrence of drugs in German sewage treatment plants and rivers. Water Res. 1998;32(11):3245–3260.

[41] Fent K, Weston AA, Caminada D. Ecotoxicology of human pharmaceuticals. Aquat. Toxicol. 2006; 7:122–159.

[42] Garric J, Ferrari B. Pharmaceuticals in aquatic ecosystems. Levels of exposure and biological effects: A review. Revue des Sciences de l'Eau/Journal of Water Science. 2005; 18(3): 307–330.

[43] Kummerer K. Drugs in the environment: emission of drugs, diagnostic aids and disinfectants into wastewater by hospitals in relation to other sources – a review. Chemosphere. 2001; 45:957–969.

[44] Miège C, Choubert JM, Ribeiro L, Eusèbe M, Coquery M. Fate of pharmaceuticals and personal care products in wastewater treatment plants – Conception of a database and first results. Environ. Pollut. 2009; 157:1721–1726.

Chapter 2

Application of Macrobiological Methods in the Settlement Wastewater Treatment

Dragan Milićević, Slaviša Trajković and Milan Gocić

Additional information is available at the end of the chapter

http://dx.doi.org/10.5772/65369

Abstract

The approach to solving the problem of water protection is changing in the world, and the opinion that wastewater is a resource instead of waste is now prevalent with research being directed in the direction of simpler, energetically more rational and more economically acceptable technological solutions for wastewater treatment, primarily in the field of biotechnology, especially there where favorable climate conditions and the use of large land areas are available. The mechanism of wastewater treatment by macrobiological methods is simple and is reduced to extraction of certain substances from wastewater directly with plants or through the food chain with animals and their concentration into macrobiological living stations. Macrobiological living stations are extracted from the water in the form of biomass by simple mechanical methods, and in that way the final removal of nutrients and other substances from the water is completed. The produced biomass can be used as food or feed, with mandatory sanitary inspection, or as an emergent in biomass production. This paper presents the principles of application of macro biologic methods in wastewater treatment and the experience gained through the research at the Faculty of Civil Engineering of Niš and at the waste water treatment facilities in Sokobanja.

Keywords: wastewater treatment, macrobiological methods, resource and energy potential

1. Introduction

The sudden technological and industrial development, tumulous demographic growth and rapid urbanization especially in the last two decades pose humanity with four big problems: water, food, energy and environment. The problem of water is especially pronounced because it is implicitly present in the remaining three problems, the production of food and energy are decisively dependent on water and the key problems of environment protection are protection of water quality and protection from the harmful effects of water.

Even today, the water crisis is well underway and according to predictions, by the middle of this century it will develop into a crisis of global proportions. The gap between available reserves of water and the increasing need for it on one and the pollution of water resources on the other hand are more and more pronounced with each day so rational use of water resources and their protection from further pollution, today and especially in the near future are developing as the main global problem.

The world today has many high technologies for wastewater treatment available, primarily of physicochemical nature, which allow for wastewater to be treated to a very high level and can satisfy all strict and more rigorous demands set regarding quality of treated effluents released into water recipients. However, the time of intensive development of these technologies was also the time of cheap energy, but today, when the evidence crisis is evident these technologies are too expensive even for the most developed countries of the world. That is why, in the world, the approach to solving the wastewater treatment problem is increasingly changing and intensive research is conducted in the direction of cheaper technologies for treatment of wastewater and protection of water from pollution.

In the last three decades, a special interest in the world is aroused by the potential of using the macrobiological methods in the waste water treatment, whose application as of natural and not artificial waste water treatment processes provide effluents of demanded quality in an economically acceptable way in technically simple objects.

2. Macrobiological methods in the wastewater treatment

Until the energy crisis which emerged in the 1970s the leading approach to secondary (biological) wastewater treatment was the philosophy of destruction of organic matter, and tertiary treatment, removal of nutrients from wastewaters and treated waters was mainly connected with complex and expensive chemical technological processes.

The hint of an energy crisis demanded directing towards cheaper technologies for treatment of wastewater and protection of water resources from eutrophication. The opinion that wastewater is a resource instead of waste became increasingly prevalent and research was intensively directed towards simpler and economically more acceptable technological solutions, primarily in the field of biotechnology especially in conditions where the use of large land areas is possible. The tendencies of questioning the philosophy of destruction of organic matter, or their mineralisation, have become more widespread, present in the previous technological practice of wastewater treatment and accepting the philosophy of synthesis of organic matter and nutrients into higher forms, which brings up the numerous matter in wastewaters [nitrogen (N), phosphorus (P), etc.] which can be interesting and useful as a raw material.

In accordance to the philosophy of the synthesis of organic matter and nutrients into higher forms, the possibility to use the macrobiological methods for the wastewater treatment attracted interest around the world.

Macrobiological methods of wastewater treatment present aerobic processes of synthesis in the direction of more complex organic matter which is easily removed from water in the form of biomass with noticeable reduction of energy needed for functioning of the system and encompass a whole series of macrobiological unit operations, the list of which, with constant research conducted around the world, continues to grow. These methods are applied as natural instead of artificial processes of wastewater treatment and they provide effluents of demanded quality in an economically acceptable manner in technically simple objects.

Intensive research in this area began in the 1970s in the world [1], and almost at the same time in Serbia [2, 3]. The starting results were obtained under the direction of Prof. Dr Lazar Ignjatović, in the period between 1975 and 1979, from the Faculty of Civil Engineering, University of Niš, Serbia and the wastewater treatment plant (WWTP) in Sokobanja, which was used as training ground for the staff of the faculty, through the project "The influence of accumulation on the change of ecosystem", with the research being continued in the period of 1978–1988, through the multidisciplinary project under the same name within the Fulbright program. In order to control the nutrient into effluent, numerous macrobiological unit operations were tested as laboratory models and then brought to the level of macro model, namely the part of an already existing wastewater treatment plant in Sokobanja, in cooperation with reputable experts from the USA for certain areas.

3. Treatment mechanism

In the wastewater treatment performed by the macrobiological methods the macrobiological living stations are used. The macrobiological living stations is the term for all the higher plants or animals with all the characteristics of living organisms, including with sexual and sexless procreation.

Living stations, either plant or animal, are mainly made of water starting from 73% in carp, 80% in terrestrial macrophytes, even up to 95% in hydrophytes. The exact percentage of water depends on the type of living station, the weight and the age of the station in the moment of sample processing.

The largest part of the living stations dry mass (usually over 90% in plants) is made of three basic elements: carbon (C), oxygen (O) and hydrogen (H), taken directly from the water or air, while a smaller part (around 10% in plants) is made of all the other elements.

The plant or animal cannot complete its life cycle in the absence of any of the necessary elements, which must be available directly or through suitable enzyme activity, provided that there are no antagonistic or toxic effects of the other elements. From 92 natural mineral known elements around 60 are found in living stations. From the 60, around 16 are considered essential for growth of plants, and approximately the same number is necessary for growth of aquatic animals of interest. It should be considered that although they do not need other elements, plants and animals accumulate some elements not significant for growth and development.

Wastewater treatment is completed through bioconcentration, or accumulation of substance from the environment and concentration in a biological station, directly in plants or indirectly, through the food chain, in animals. Because living matter is formed by a few biogenic elements either plastic: hydrogen (H), oxygen (O), carbon (C) and nitrogen (N) or oligo elements: zinc (Zn), copper (Cu), iron (Fe) and magnesium (Mg) in the presence of phosphorus (P), natrium (Na) and potassium (K), chloride (Cl) and manganese (Mn) the factor of bioconcentration of the substances from the environment in the organism is important along with the dynamic of the process, or the rate of bioconcentration of the substance of interest through time.

Numerous factors affect the growth, development and reproduction of the living stations. Some of them are crucial while others are less significant, depending on the phase of growth of the living station. The earliest developmental phases are the most sensitive in all living stations. Light and temperature, along with activational energies have crucial effect on the dynamic of the process. Plant species of areas with temperate continental climate are active when the temperature of the water reaches above 15–17°C, while tropical species are mainly active from 20 to 24°C.

If the natural conditions of the environment are favorable, application of macrobiological stations is possible in natural conditions, perfectly cheap because free energy of the Sun is used as an energy source, and the water itself serves as a collector with the least loss.

If the natural conditions of the environment are unfavorable, the influential factors can be put under control and then we have artificial environmental conditions. Application in green-houses in periods of unfavorable conditions of the environment, with the addition of light and thermal energy, is possible. In those cases great economic effects of use in natural conditions decrease. There where geothermal energy is available, its use may be rational for extended work over the whole year.

It is clear that the wastewater treatment mechanism is simple. It consists of taking certain substances from the water (whereby the water is rid of this substance) and its bioconcentration into microbiological living stations. By removing larger macrobiological living stations in the form of biomass from the water with simple mechanical methods the nutrients and other substances are finally removed from the water.

The final disposition of biomass, depending on its nature, is performed by the standard transport means. If the biomass is used as a nutrient or food (with necessary sanitary control) it has market value which considerably exceeds the transportation costs.

It is a highly clean technology in wastewater treatment using clean energy (solar and/or geothermal) with the final product being usable biomass. The civil engineering objects are usually made of soil, relatively simple and followed by a minimal equipment fond, which significantly influences low investment costs. In investment costs the land makes as significant item but if there is unsuitable agricultural or commercial land close to the settlement available, ideal conditions for the application of these technologies are acquired.

4. Macrobiological living stations

For a living being to qualify as a macrobiological living station, respectively, a technological element in wastewater or sludge treatment it has to satisfy special criteria:

- It has to belong to a fast growing species with a short reproduction cycle, so the processes can run in an accelerated speed.
- It has to be easily removed from objects in which it is used with the purpose of wastewater treatment, namely it has to allow low manipulation costs with the produced biomass and the biomass should have value of use, which in turn affects the relief of wastewater treatment cost.

Great attention should be given to the question if the macrobiological living station can, for a longer period of time, survive and reproduce in natural habitat on its own and if the species is invasive. This is of great importance, from the aspect of possible ecological effects if a macrobiological living station finds its way into the natural environment, out of the object where it is used under control.

On the basis of the mentioned criteria for the macrobiological unit operations, only a small number of plant and animal species can be qualified. They predominantly originate in the tropical zones.

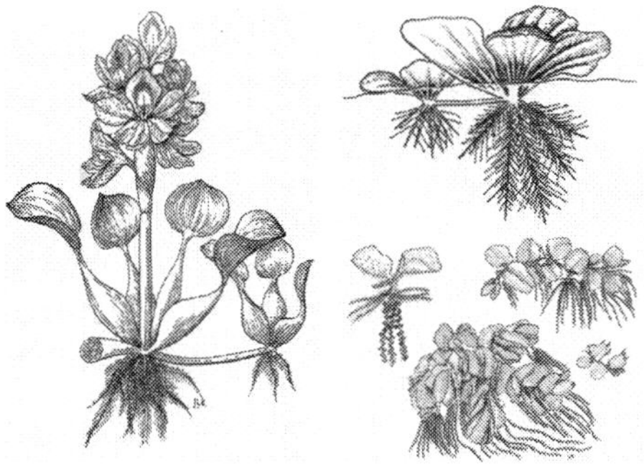

Figure 1. Floating macrophytes: *Eichornia crassipes, Pistia stratiotes* and *Salvinia* [3].

Macrobiological living stations, which qualify as a technological element in wastewater treatment technology, are classified by groups that are floating macrophytes, fish, mussels, earthworms, etc. The list of macrobiological living stations is very wide, but it does not encompass all possibilities because in this area intensive research is present.

The list of possible floating macrophytes should be made of hydrophytes without woody tissues, especially ones which float on the surface of the water. These plants cannot adapt to the change of the water level so for their normal growth and development the water level must be kept approximately constant.

The representatives of this plant group use food directly in the shape of dissolved nutrients in the water. Some do that only through the leaves, e.g., floating crystalwort (*Riccia fluitans L.*) to some extent lemna (*Lemna minor L.* and *Lemna trisulca L.*), while other especially larger species do that through their roots which hang in the water, e.g., water hyacinth (*Eichhornia crassipes Martius*), water lettuce (*Pistia stratiotes L*) and salvinia (*Salvinia natans L.* and *Salvinia auriculata Aublet*) (**Figure 1**).

In terms of fish, fast-growing species capable of consuming large quantities of food are of interest. Also of interest are food pyramids because of the choice of fish species, especially because of interrelationships in polyculture composition. Phytophagous species have a special role because they lean directly on the primary production in the aquatorium in the food chain.

Figure 2 shows a food chain and fish species of predominant interest: silver carp (*Hypophthalmichthys molitrix*), grass carp (*Ctenopharyngodon idella*), bighead carp (*Hypophthalmichthys nobilis*) and common carp (*Cyprinus carpio*). The list can be expanded with some tropical fish species, e.g., tilapia (*Tilapia aurea*) and thai catfish (*Clarias batrachus Linnaeus*). The use of tropical fish species applies the same demands as the use of tropical plants.

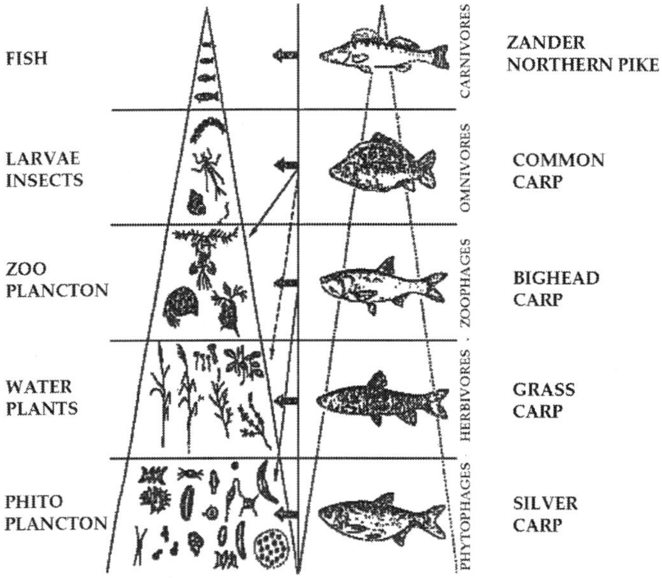

Figure 2. Fish-food pyramid and the primary interest species (based on Ref. [3]).

Fish, as poikilothermic animals, because of poor adjustment to sudden temperature changes, must not be rapidly transferred from one water environment to the other if the water temperature difference is greater than 2°C, because this leads to temperature shock and death in most species. The fish must be transferred carefully, because being thrown in the water during transfer leads to bursting of the swim bladder.

The role of mussels in removal of suspended and colloid material from wastewater deserves great attention from the researchers. It was experimented only with one species of mussel from the temperate climate belt in the Faculty of Civil Engineering of Niš. It was experimented with the zebra mussel (*Dreissena polymorpha*) (**Figure 3**). It should be noted that this species of mussel may pose not only an ecological threat, but also great danger to the hydrotechnical systems and objects, so it should be used in strictly closed systems.

Figure 3. Mussels – *Dreissena polymorpha* [4], *Unio pictorum* [5] and *Anodonta cygnea* [6].

Species of interest would also probably be the painter's mussel (*Unio pictorum*), swan mussel (*Anodonta cygnea*) (**Figure 3**), Eastern Asiatic freshwater clam (*Anodonta (Sinanodonta) woodiana*) and large far eastern mussel species (order *Cristaria*) whose shell can grow over 30 cm. This area is open for research with warning if nonendemic species are in question.

Two species of earthworms are of interest as macrobiological living stations in the technology of sludge treatment: red Californian earthworm (*Eisenia fetida*) and red earthworm (*Lumbricus rubellus*) (**Figure 4**). For further research the red tiger earthworm (*Eisenia andrei*) (**Figure 4**) is interesting because it can treat rich organic waste in massive amounts. Probably the European nightcrawler (*Eisenia hortensis* or *Dendrobaena veneta*) would also be of interest but it is considered an invasive species which should be used in strictly controlled conditions without being allowed into the natural environment [7].

Figure 4. Earthworms – *Eisenia fetida* [8], *Lumbricus rubellus* [9] and *Eisenia andrei* [10].

For the needs of hydroponics unit terrestrial plants are used, such as tomatoes, e.g., the American flowerpot tomatoes (*Licopersicum esculentum*), leafy vegetables, e.g., chard (*Beta vulgaris*), corn for silage and similar species. The list is very long because numerous fruits or vegetables can be used.

Carefully composed polycultures (bigger number of different species of macrobiological living stations in a unique aquatorium) have a bigger effect on the quality of effluents than a monoculture (a single species of a microbiological living station in a unique aquatorium). The reason being that a monoculture drains a narrow circle of substances and because of that it has limited effect in removing nutrients and wastewater treatment. The advantage is given to monocultures only in the case of final biomass derivation if it is used for human or animal consumption or partial wastewater treatment.

If ambient conditions favor some of the members of the polyculture, it spontaneously comes to suppression of the other members and the formation of a monoculture, namely population of a macrobiological living station which ambient and other factors provide the most suitable conditions. In those cases instead of insisting on polycultures the transition on a series of monoculture basins is expedient.

It should be mentioned that successive application of monoculture basins enlarges the investment costs. But continual additional introduction of macrobiological living stations from external sources, for polyculture maintenance, is usually more expensive than amortization of bigger investment costs in more basins.

In all cases parent clusters under optimal conditions must be ensured. This is optimally in the shape of a macrobiological living station bank on a regional level, for example botanical gardens or zoos, organized on a wider administrative area.

5. Objects and system design

Working on the choice of unit operations, their synthesis into the technological process and objects and system design is complex engineering work which demands professional experience along with team work of participants of the system design. Designing objects and the system for wastewater treatment starts from the available information on the wastewater quality, defining the type and concentration of the unwanted substances and the needed removable level. Based on the analysis, the technological scheme of the wastewater treatment system is defined.

Based on information on the amount of wastewater and its variation, hydraulic and process loads are defined. If the variations are big, the problem of synchronized hydraulic and process loads must be solved by choosing adequate modular object units. In this phase, decisions are made on the choice of macrobiological unit operations and the choice is made between mono and polycultures.

It should be kept in mind that for synthesis of macrobiological unit operations into the treatment processes, aside from macrobiological, standard (classic) unit operations are often incorporated with the purpose of bringing characteristics of wastewaters on the effluent of wanted quality. Although any wanted level of wastewater quality may be achieved through a planned combination of unit operations, the choice is made under clear economical conditions.

As with any modern biological system of wastewater treatment, primary treatment must always precede a system based on macrobiological methods. For primary treatment of wastewater the use of a highly efficient (tubular) settlement tank from which the primary sludge is processed by anaerobic decomposition in digesters and on vermiculture (VF) fields next to smaller wastewater treatment plants is recommended.

When secondary treatment is in question, unlike classic technologies with microbiological population with which secondary treatment is made of a microbiological unit and a secondary settlement tank, with macrobiological methods there is no need for a secondary settlement tank. The reason being the lack of secondary (biological) sludge because the transformations of materials from wastewaters, through the food chain, are done into the biomass of the macrobiological living station.

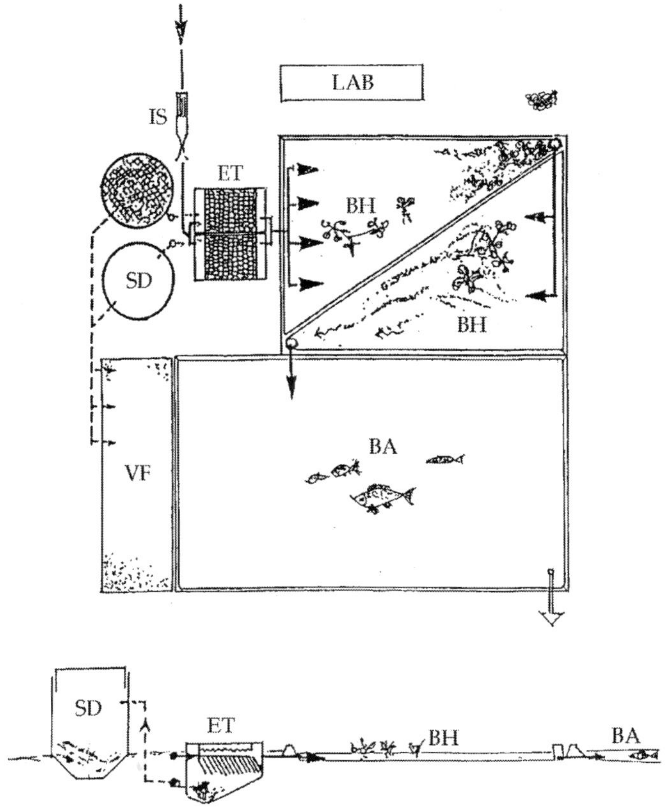

Figure 5. Scheme of the human settlement waste water treatment facility [3, 11]. IS – inlet structure; ET – efficient settlement tank; BH – basin for sanitary hydrophytocultures; BA – basin for sanitary aquacultures; SD – sludge digester; VF – vermiculture fields.

Based on the technological scheme, after choosing unit operations and defining modular units, the technological scheme of the system with the basic hydraulic and technological calculations is designed. This results in a horizontal plan of the objects and their height scheme from the entrance to the exit of the treatment plant.

After the place and the role of some objects, their sizes and height positions are defined, the design of the objects for application of macrobiological unit operations is reduced to civil engineering design of objects. For object design, the knowledge of unit of macrobiological living station, design information and characteristic technical details of the object is required. Knowledge of civil engineering design, stability and dimension of constructions and civil engineering regulations is also required.

For application of macrobiological unit operations, two tendencies are present:

- For smaller agglomerations, especially with seasonal problems, macrobiological unit operations are synthesized into complex, with cheaper investment and maintenance objects.
- For bigger agglomerations behind classical treatment plants these methods are used for polishing of effluent quality with nutrient removal (tertial treatment).

The example for the first approach for smaller settlements is given in the follow-up. The scheme of a wastewater treatment plant is given in **Figure 5** with object marks. Each of the objects is described in more detail with needed design information.

The shown scheme is applicable for settlements without industrial and toxic wastewater. The scheme incorporates wastewater treatment and sludge stabilization so that they can be disponated into the natural environment without negative ecological effects behind the treatment plant. This is ecologically clean technology.

Depending on the ability and concert of the operator the removal of suspended solids is from 80% to above 95%, and this applies for putrescible matter too. The reduction of bacteria is above 99% so the water can be used for irrigation in semiarid areas without danger.

This technological scheme is more favorable in the level of efficiency and the produced biomass if the climate conditions are warmer and insolation is more intensive. In areas of temperate continental climate the starting hypothesis is the disposal of solar energy during the whole year or geothermal energy, if continued work of the system with low cost investment and maintenance is desired during the whole year.

The technological scheme and all unit operations are checked on the wastewater treatment plant in Sokobanja, Serbia which served as a pilot treatment plant with the process scale of 1:1, under realistic conditions. In the follow-up, description and instructions for some objects are given.

5.1. IS – inlet structure

Inlet structure (**Figure 6**) serves for removal of large suspended matter and measurement of flow of wastewater. It is made of a channel with a grid which continues to the Parshall flume. The space between iron flat bars is 2–5 cm. The slope is 1:2 to 1:3 for easier cleaning. For smaller treatment plants the cleaning is done manually with loading of handcarts and daily transport to burial of the material from the grid to a suitable location in the treatment plant area. For bigger treatment plants the cleaning of the grid is automatic and the transport of material is off the grid to the landfill.

Figure 6. Inlet structure and efficient settlement tank on the WWTP Sokobanja.

Based on the known, defined hydraulic load, standard hydraulic calculation of width of the channel for defined level and loss is completed.

5.2. ET – efficient settlement tank

The primary settlement tank is based on the system of a highly efficient tubular settlement tank (**Figure 6**) which includes a separator of oil and grease into compact construction.

Domestic wastewater treated on this type of settlement tank with a process load not greater than 0.6 l/s po m^2 (horizontal area of settlement tank) is of such quality that without further treatment it can go on macrobiological units. Water is kept in the settlement tank shortly, 15–25 minutes and there is no danger of transit into septic state which is of great importance for the effluent quality. The sludge from the settlement tank is pumped into the sludge digester (SD) for further treatment.

Based on the known, defined hydraulic load, the calculation is completed by standard procedure for primary, mechanical wastewater treatment.

In the case of different primary treatment or no treatment, the quality of wastewater should be brought to an acceptable one for macrobiological living stations which will be used in further treatment.

5.3. BH – basin for sanitary hydrophytocultures

In this basin (**Figure 7**), the dissolved and colloid matter from the wastewater is transformed into biomass of floating macrophytes under the influence of solar energy. Basin depth of 0.4–

0.6 m with a protective bank or edge of 0.2 m above the water level is recommended. The insertion of young macrophytes can be done with monocultures or polycultures depending on whether there is previous experience with the wastewater being treated. In highly polluted wastewater, *Eichhornia crassipes* is the most active species; in medium-polluted wastewaters, *Pistia stratiotes* should be given advantage and in the least polluted wastewaters, *Salvinia* is most appropriate. The quality of the wastewater, the choice of macrobiological living station and process loads define the dimensions of the basin. For domestic wastewater and daily specific consumption of 250 l/person from 3 to 5 m^2/PE of area under the hydrophytoculture *Eichhornia crassipes* is needed, while *Pistia stratiotes* demands double of that value.

Figure 7. Basin for sanitary hydrophytocultures and basin for sanitary aquacultures on the WWTP Sokobanja.

The growth dynamic of green biomass of floating macrophytes and area coverage of the basin in green biomass is in function of plant quality, insolation and temperature. For temperate climate conditions, based on the research done in the wastewater treatment plant in Sokobanja, the basin area coverage for the *Pistia stratiotes* biomass is 2–25 kg/m^2 and for *Eichhornia crassipes* biomass it is 5–35 kg/m^2 [3, 12]. The growth dynamic of green biomass of larger floating macrophytes is well presented by the exponential equation $B = B_0 \cdot e^{kt}$ in which B is the probable green biomass after a certain time t in kg, B_0 is initial green biomass in kg, k is the rate of growth in 1 day and t is time in days [3, 12]. The rates of growth for *Eichhornia crassipes* are 0.130 for 30°C, 0.052 for 20°C and 0.015 for 15°C and for *Pistia stratiotes* 0.061 for 30°C, 0.026 for 20°C and 0.010 for 15°C [3, 12].

For basin coverage with green biomass of 20 kg/m^2 and average temperature of 20°C, daily wet biomass growth about 5 t/ha for *Pistia stratiotes* and about 10 t/ha for *Eichhornia crassipes*. Biomass extraction in smaller treatment plants is manual. The biomass dries on a bank if fresh, green biomass is not used. Mechanized extraction of biomass is possible and economically justified in bigger treatment plants.

The work of this part of the treatment plant is connected to temperature and insolation conditions and under natural climate conditions in areas of temperate continental climate is possible in the period between May and October. In the case of continued process during the year greenhouses and introduction of additional thermal and solar energy are needed.

5.4. BA – basin for sanitary aquacultures

Under the effect of solar energy, through primary production the process of nutrient removal and transformation into the biomass of fish is completed. Phytoplankton and zooplankton which have used the nutrients are transformed through the food chain of herbivore and carnivore fish into a high value protein.

Average basin depth of 0.3–0.7 m is recommended (**Figure 7**). Young fish are inserted into the basin in spring, most often with polycultures of herbivore fish with the addition of carp. Depending on the input water quality, dimensioning of the basin and aquapolyculture composition are completed.

If primary water treatment is done through tubular settlement tanks by the usual specific consumption, hydraulic load of $Q \leq 50\ m^3/ha$ for a day is recommended. Stocking is done with 200–400 kg/ha by polyculture (silver carp 65–50%, bighead carp 22–30%, grass carp 8–10% and river and ponds common carp 5–10%).

Production of biomass for a season depends on the success of plant management, and it ranges from 1.200 to 2.000 kg/ha for a season of 200 days.

5.5. SD – sludge digester

Sludge digestion (**Figure 8**) is desirable for hygienic and esthetic reasons, although it is not necessary if macrobiological treatment of the sludge is completed.

Figure 8. Sludge digester and vermiculture field on the WWTP Sokobanja.

The amount of sludge and digester dimensions is calculated by the process for classical problem solutions. The calculation of sludge pumps and pipes, with notice that the pipes should not be under Ø 200 mm because of sludge flow resistance, is done by standard procedure. The same applies for the use of biogas.

If hydrophytocultures are used as an emergent for biogas production, the volume of the digester is to be increased by 20%. Construction of a lateral opening with a nonreturn flap for insertion of the biomass into the digester is mandatory, along with a spiral access ramp or a lift.

5.6. VFs – vermiculture fields

On the vermiculture field (**Figure 8**) sludge treatment into highly valued hummus is completed, using *Eisenia fetida* or *Lumbricus rubellus* living stations.

If fresh sludge was put on vermiculture fields, no matter the addition of wood chips or cut paper for moisture reduction, odor and insects may appear.

The excess of earthworms is returned to the basin for sanitary aquacultures and can be used as food for the fish with a goal of increasing their growth.

Turning plant biomass into hummus if by far more rational than gasification and it gives a valuable commercial product, it especially increases the quality of the total produced material.

Macrobiological unit operations are especially important for a greater number of smaller agglomerations, in which rational expansion of wastewater treatment systems is possible so they give an effluent of high quality. The use of macrobiological unit operations, along with other classic operations in the technological scheme, allows not only a cheap but a technically simple and safe solution to nutrient (nitrogen, phosphorus) removal and BOD5 reduction from wastewater of settlements without industrial wastewater.

6. Usage of the resource and energy potential of waste waters

The macrobiological living stations use from the waste water the nutrients and other elements which are a part of the biomass for their growth and development. The produced biomass has a practical value, so the nutrients and other matter in the waste water are not only harmful matter to be removed from the waste water, but are also the resource for production of the biomass. Regarding that the macrobiological living stations should belong to the fast-growing species with the short reproductive cycle, the quantity of biomass produced in the waste water treatment process, are huge, as a rule.

Based on previous research some possibilities for the use of biomass were noticed, but this is an area which is yet to be thoroughly researched.

Floating macrophytes can be widely used as biomass [3], especially when water hyacinth (*Eichornia crassipes*) and water lettuce (*Pistia stratiotes*) are in question [13–15].

Water hyacinth and water lettuce mixed with sludge are great material for hummus production, especially for winter cover and thermal protection of the vermiculture.

Water hyacinth combined and composted with manure gives better quality material for vermiculture nutrition with acceleration of the population dynamic because the root system is the ideal habitat for laying cocoons and reproduction.

Water hyacinth is ideal for nutrition of nutria (coypu) and they would rather be fed the hyacinth than beet leaves. Detailed information about the possibilities of application of floating macrophytes in livestock keeping, including its possible application in silage, is not available.

Fish as biomass can be used as food in human nutrition, especially if higher quality species are in question: grass carp (*Ctenopharyngodon idella*), silver carp (*Hypophthalmichthys molitrix*), bighead carp (*Aristichthys nobilis*), walking catfish (*Clarias batrachus*) and common carp (*Cyprinus carpio*). The possible danger from bioconcentration of heavy metals and pesticide, as well as quarantine in connection with epidemiological control of food quality should be mentioned.

Mussel meat, which is easily removed with hot water, is rich in proteins and is eaten by poultry and pigs in dried form. The shell of some mussels, e.g., painter's mussel, can be used as material for nacre products.

Worms, both species: red Californian earthworms (*Eisenia fetida*) and red earthworms (*Lumbricus rubellus*) which are interesting as macrobiological living stations in the sludge and manure treatment technology can be widely used as a biomass.

Earthworms can directly be used as food for poultry, pigs and pets (birds, tortoises, iguanas, snakes, fish in aquariums). They can also be used for production of protein flour, a high quality protein component in dry condition which is added to fish flour or feed. Commercially, the use of earthworms for the nutrition of aquacultures of fish is very favorable, especially for cultivation and fattening of fastgrowing Clarias catfish and cultivation of trout.

Even more profitable is placement of earthworms as bait (in fishing or for attracting wild birds) through specialized stores for hunter and anglers.

Earthworms feed on detritus, decomposing organic matter, and as secretion finely crushed material appears, relatively stable vermicompost, namely humus. Humus has great value as a natural fertilizer because it improves the structure of the ground and reduces/eliminates the need for chemical fertilizers. It is used in flower and vegetable cultivation, for nursery gardens, orchards and lawns, for topsoiling of surfaces or as a component in soil devastated because of the use of chemical fertilizers. If there was no pesticide, heavy metals or toxic substances in the starting material, the humus can be used in production of healthy food because, aside from having a positive effect on the ground, it has a positive effect on various plants and crop plants.

Regarding that, when larger floating macrophytes are in question, the amounts of biomass produced daily in the process of wastewater treatment are quite large, the possibility of using said biomass as raw material for production of biogas is of great importance. The produced biogas can further be used for combined production of electrical and heat energy.

Combined production of electrical and heat energy [combined heat and power (CHP)], also called cogeneration, is the production of electrical power out of the natural gas, biogas and waste matter disposal site gas, with the simultaneous usage of waste heat which is otherwise lost in the industrial process. Modern cogeneration systems today reach efficiency above 90%, that is why cogeneration presents the most efficient and economically most justified way of reducing high energy costs in industrial plants and municipal objects.

In wastewater treatment plants with an anaerobic reactor (digester) for sludge stabilization biogas occurs as a mixture of combustible and noncombustible gases with the average composition of (in cubic %): methane 55–75%, carbon-dioxide 25–45%, other gases like hydrogen, oxygen, carbon-monoxide, nitrogen, hydrogen-sulfide, ammonia and water vapor [16]. The efficiency of biogas production is provided by maintaining temperature, pH value, by mixing and removal of oxygen and toxic matter.

Production of biogas can be assessed based on the following practical and experimental information [16]:

- on municipal wastewater treatment plants the average production of biogas 25 l/PE per day;
- with industrial wastewater (sugar refineries, molasses processing, potato processing, fruit juice production, dairy farms, breweries, paper and cellulose) the average methane production is 0.20–0.30 m^3/kg CSB with the methane fraction in biogas being 60–80%.

The heat power of biogas depends on the methane content and for the average content of 65% methane it is equal to 6.4 kWh/m^3 [16]. That is how it is possible to produce 2.5 kWh of electrical and 3.3 kWh of heat energy from 1 m^3 of biogas with the reduction of CO_2 emission above 50% in a practical operation on a cogeneration plant with gas motors (**Figure 9**) [16].

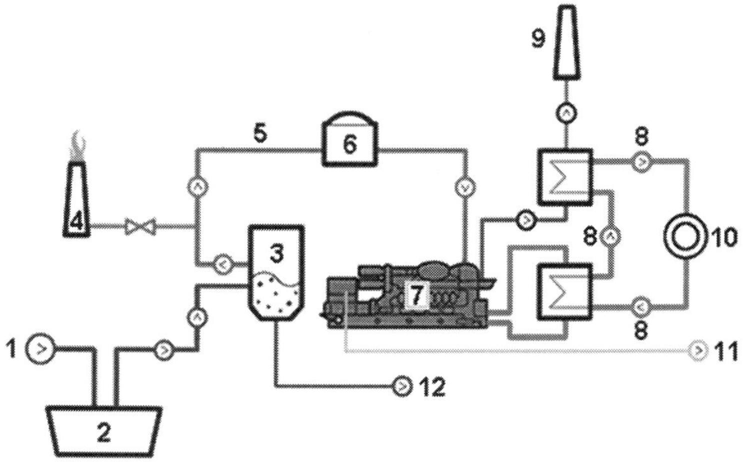

Figure 9. Usage of biogas at the facilities for waste water treatment (based on Ref. [17]). 1 – biomass; 2 – sludge thickening; 3 – anaerobic digester; 4 – gas torch; 5 – biogas; 6 – gas tank; 7 – gas engine; 8 – heat exchanger; 9 – exhaust; 10 – heat energy; 11 – electrical energy; 12 – agricultural fertilizers.

In wastewater treatment plants with a basin for sanitary phytocultures, the produced biomass of floating macrophytes, either processed through a digester for biogas production increase, used in cogeneration plants with a gas motor, or directly burned in cogeneration plants with an indirect gas turbine process, the amounts of produced electrical or heat energy can be multiply increased relative to plants with classic technologies, which of course directly depends from the available basin area for sanitary hydrophytocultures and the daily growth of the biomass of floating macrophytes.

The dry mass of *Pistia stratiotes* is 4.9% and for *Eichhornia crassipes* 4.6% from the green mass for the leaf part of the plant (the variations of the root mass are great). Based on the literature information each kilogram of *Eichhornia crassipes* dry mass gives 370 l of biogas, whose heating value is around 6.1 kWh/m^3 [13]. For *Pistia stratiotes,* keep in mind that the structure of biomass is similar to the previously mentioned plant.

In **Table 1**, the values of biomass growth, biogas amount and electrical and heat energy, which can be produced from the biogas are shown for *Pistia stratiotes* and *Eichhornia crassipes,* calculated based on the previously stated experimental data. All values are given in ha of basin area under sanitary hydrophytocultures by day.

	Air temperature (°C)	*Pistia stratiotes*			*Eichhornia crassipes*		
		Area coverage (kg/m²)			Area coverage (kg/m²)		
		2	20	25	5	20	35
Green biomass growth (t/ha by day)	15	0.201	2.010	2.513	0.756	3.023	5.290
	20	0.527	5.268	6.585	2.669	10.675	18.682
	30	1.258	12.580	15.725	6.941	27.766	48.590
Dry biomass growth (t/ha byday)	15	0.010	0.098	0.123	0.035	0.139	0.243
	20	0.026	0.258	0.323	0.123	0.491	0.859
	30	0.062	0.616	0.771	0.319	1.277	2.235
Biogas amount (m³/ha by day)	15	3.644	36.442	45.552	12.861	51.445	90.029
	20	9.551	95.512	119.390	45.423	181.691	317.959
	30	22.807	228.071	285.089	118.143	472.572	827.001
Electrical energy amount (KWh/ha by day)	15	9.110	91.105	113.881	32.153	128.612	225.071
	20	23.878	238.781	298.476	113.557	454.228	794.898
	30	57.018	570.179	712.723	295.357	1181.430	2067.502
Heat energy amount (KWh/ha by day)	15	12.026	120.258	150.323	42.442	169.768	297.094
	20	31.519	315.191	393.988	149.895	599.580	1049.266
	30	75.264	752.636	940.795	389.872	1559.487	2729.102

Table 1. Biomass growth, biogas amount and energies which can be produced for larger floating macrophytes [18].

As it may be concluded on the basis of the displayed values, at the facilities for waste water treatment with the basin for sanitary hydrophytocultures, significant quantities of electric and thermal energy can be obtained through the cogeneration.

Part of the produced electrical energy can be used for the plants' own needs, and extras can be forwarded into the ED network, while the produced heat energy can be used for maintaining the temperature in the digester, to ensure the efficiency of biogas production. The heat energy can also be used for providing favorable conditions (air temperature from min. 20°C and area coverage from min. 20 kg/m^2) in greenhouses for application of these technologies in our climate conditions during the whole year, which in turn provides constant growth of floating macrophyte biomass and annuls the seasonal character of macrobiological methods.

Considering the global climate changes, the Kyoto protocol predicts the possibility that developed countries invest in modernization of industrial and energy power plants and reduction of carbon dioxide emission and other gases which cause the greenhouse effect on the territories of undeveloped and developing countries. As introduction of cogeneration in wastewater treatment plants by macrobiological methods and the usage of surpluses of electrical and heat energy in the energy system is in accordance with the Kyoto protocol, it would allow receiving of exceptionally favorable credits and investments for energy and ecology sector, and is as such of great importance, especially for developing countries, which are yet to solve the problem of settlement and industrial wastewater and the building of plants for their treatment.

Considering that the yields, which amount to a few dozen tons by ha daily with larger floating macrophytes in favorable insolation and temperature conditions, are of fantastic size , research of their value of use in animal husbandry, and even more in energetics is extremely significant.

7. Conclusion

Based on previously achieved results, it is evident that macrobiological unit operations will in the future find their place in the technology of wastewater treatment for multiple reasons:

- instead of destruction of material, namely the stopping of natural processes we are going towards the philosophy of synthesis of organic matter into higher levels of biomass;
- the processes of synthesis use natural energy sources (sun, heat) and on that basis they present ecologically highly "clean" technologies;
- the objects are relatively simple civil engineering objects made of land followed by a minimal equipment fond, which greatly impacts low investment costs;
- there is no biological sludge and no parts of the object which represent the secondary settlement tank and sludge line, which significantly affects the relief of wastewater treatment costs;

- the final product is biomass which has value of use as food or feed, or as an emergent in biogas production, which in turn affects the reduction, even complete annulment of wastewater treatment costs.

The energy crisis, which is deepening day by day, high prices of energy, materials and the workforce, the demand for low investment and operation costs, and the sharper requests set in regards of discharge treated wastewater into recipients make the application of macrobiological methods in wastewater treatment around the world even today, and especially in the near future come to the fore and intensify the research in this extremely important area.

Author details

Dragan Milićević*, Slaviša Trajković and Milan Gocić

*Address all correspondence to: drgara@gaf.ni.ac.rs

The Faculty of Civil Engineering and Architecture of the University of Niš, Aleksandra Medvedeva, Niš, Serbia

References

[1] Stewart K.K. Nutrient removal potentials of various aquatic plants, Hyacinth Control J.; 1970; 8:34-35

[2] Ignjatović L. Industrial wastewater treatment, Part V - Macrobiological methods in wastewater treatment. In: Miloje Milojevic, editor. Civil Engineering Calendar 1993. Association of Civil Engineers and Technicians of Yugoslavia, Belgrade; 1992. p. 334-395. (in Serbian)

[3] Ignjatović L. Microbiological methods in wastewater treatment - Guide through the unit operations. ECO-TECH, Belgrade; 1995. 222 p. (in Serbian)

[4] The zebra mussel [Internet]. Available from: http://www.ebrenavegable.cat/musclo-zebrat.php

[5] Freshwater mussels [Internet]. Available from: http://www.gimnazijaso.edu.rs/gornjepodunavlje/zivotinje/beskicmenjaci/mekusci/skoljke.php

[6] The Naiadi garden. Freshwater bivalves: Natural filters [Internet]. Available from: http://giardinonaiadi.blogspot.gr/2012/08/bivalvi-dacqua-dolce-filtri-naturali.html

[7] What are the earthworms suitable for vermicomposting [Internet]. Available from: http://www.lumbri.com.hr/hr/gliste/koje?su?gliste-pogodne-za-vermikompostiranje/

[8] File: Eisenia foetida R.H. (2).JPG [Internet]. Available from: https://commons.wikimedia.org/wiki/File:Eisenia_foetida_R.H._(2).JPG

[9] Discover Life. Lumbricus rubellus Hoffmeister, 1843. [Internet]. Available from: http://www.discoverlife.org/mp/20q?search=Lumbricus+rubellus

[10] The Composter Newsletter. Wriggle your way into my heart! [Internet]. Available from: http://archive.constantcontact.com/fs057/1101915097096/archive/1102821409563.html

[11] Milićević D., Milenković S., Nikolić V. Application of macrobiological methods in settlements wastewater treatment, Water and Sanitation Equipment; 2006; 1:31–37 (in Serbian)

[12] Milićević D. Modeling system to use nutrients from the water using macrophytes [Unpublished master thesis]. Civil Engineering faculty of Nis, Serbia; 1994. (in Serbian)

[13] Center for New Crops & Plant Products, at Purdue University. Eichornia crassipes (Mart.) Solms [Internet]. Available from: https://hort.purdue.edu/newcrop/duke_energy/ Eichornia_crassipes.html

[14] Lareo L., Bressani R. Possible utilization of the water hyacinth in nutrition and industry [Internet]. Available from: http://archive.unu.edu/unupress/food/8F044e/8F044E0c.htm

[15] Sanmuga Priya E., Senthamil Selvan P. Water hyacinth (Eichhornia crassipes) – An efficient and economic adsorbent for textile effluent treatment – A review. Arabian Journal of Chemistry, Available online 13 March 2014 [Internet]. Available from: http://www.sciencedirect.com/science/article/pii/S1878535214000562

[16] MPD TRADE, Belgrade. Production and characteristics of biogas [Internet]. Available from: http://www.kogeneracija.rs/biogas.html

[17] MPD TRADE, Belgrade. The emergence and use of biogas [Internet]. Available from: http://www.biogas.rs/nastanak.html

[18] Milenkovic S., Milicevic D., Nikolic V. Application of macro biologic methods in wastewater treatment and usage of its energy and resource potential. International Conference on Water Observation And Information System For Decision Support BALWOIS, Ohrid, Republic of Macedonia, 23–26 May 20066

Chapter 3

Microbe-Based Strategy for Plant Nutrient Management

Shaon Ray Chaudhuri, Madhusmita Mishra, Sonakshi De, Biswajit Samal, Amrita Saha, Srimoyee Banerjee, Abhinandan Chakraborty, Antara Chakraborty, Sonali Pardhiya, Deepak Gola, Joyeeta Chakraborty, Sourav Ghosh, Kamlesh Jangid, Indranil Mukherjee, Mathummal Sudarshan, Rajib Nath and Ashoke Ranjan Thakur

Additional information is available at the end of the chapter

http://dx.doi.org/10.5772/67307

Abstract

The rapid industrialization and urbanization of developing countries such as India have encroached on cultivable lands to meet the demands of an ever-increasing population. The altered land use patterns with increased fertilizer use has increased crop yields with leaching of major portion of the applied nutrients from the soil. Nitrates and phosphates are the agricultural pollutants that are discharged into aquifers due to anthropogenic reasons causing severe environmental and health problems. Production of these nutrients requires energy and finite resources (rock phosphate, which has gradually depleting reserves). An alternative management strategy would be to sequester excess nutrients within a biomass that is reused for agriculture. Two discrete enriched microbial consortia with the potential of simultaneous nitrate and phosphate sequestration upon application as biofertilizer restricted them within the plant root zone, ensuring prevention of eutrophication through leaching while making it available for uptake by plants. The nutrient accumulated biomass enhanced the crop yield by 21.88% during mung bean cultivation with maintained elemental content and other nutritional qualities. The major drawback of conventional biofertilizer application (slow release and action) could be overcome using this formulation leading to environmental protection, crop yield enhancement and soil fertility maintenance post-cultivation.

Keywords: nitrate accumulation, plant growth promotion, phosphate accumulation, phosphatase activity, microbial consortium

1. Introduction

In developing countries like India, rapid industrialization and urbanization have led to encroachment of cultivable lands. The agricultural practices are being gradually modified to increase the food production so as to meet the need of the ever-increasing population. The significant increase in the use of inorganic and organic fertilizers as well as alterations in the land use pattern has led to high yield of crops. But the major disadvantage that emerged out of such practices is the gradual leaching of nutrients and harmful chemicals in the soil and water. Nitrate is one such common agricultural pollutant discharged into the aquifers. Other potential sources of nitrate are the geological processes like eruptions, flood and land silting, irregular rainfall and stream flow patterns, natural process of plant decay and organic residues, anthropogenic sources of land practices, traditional agricultural practices like dry farming, marginal irrigation, large scale flood plain farming and application of fertilizers, leaching from paddy and tea cultivation, sewage infiltration, reuse of agricultural land for human settlement, industrial chemical spills and landfill leachates [1–10]. Nitrate pollution has thus emerged as a global problem and happens to be the second most dangerous pollutant after the pesticides [11, 12]. In marine environment, it induces plankton bloom destroying the native flora and fauna of the region [13]. In humans, it causes condition known as methemoglobinemia (blue baby syndrome) in infants and disorders of central nervous system, cardiovascular system as well as gastrointestinal system while posing to be carcinogenic [14].

The permissible nitrate level in ground water (10 mg/l for NO_3–N and 45 mg/l for NO_3) has been demarcated by "United States Environmental Protection Agency (EPA)." Some of the conventional methods for nitrate removal from water include distillation, reverse osmosis and ion exchange. These processes are quite complex as well as expensive which limits their application during scale up of processes. Bioremediation appears as a desired alternative [15–17], but the major limitation for its application is the longer retention time as compared to the physicochemical processes. Lately the membrane technology of denitrification has been blended with biological immobilization techniques to achieve efficient operation. This combination helps minimize the associated problem while making the process economically viable [18]. Electro bioremediation where effect of electric field is observed on pollutant reduction has also been studied [19–21]. Nitrate reduction by biological means has been reported to be carried out in fluidized expanded bed bioreactors [22], submerged membrane bioreactor [23], continuous flow bioreactors [24] as well as packed bed reactor [25] with PVS tubes [26], alginate [27], K- Carrageenan [28] and microbial cellulose [29] as immobilization matrices. It could either be through assimilatory or dissimilatory pathway. An alternative pathway of nitrate removal is through nitrate accumulation as evident in Isolates of **genus** Beggiatoa, Thiomargarita and Thioploca, as well as one species of Bacillus [30].

Phosphate is another essential plant growth nutrient which is lost in wastewater from domestic, industrial (dairy as well as detergent) and agricultural sectors [31]. It also causes eutrophication upon seepage into the surface and ground water bodies. Phosphate is derived from rock phosphate whose reserves are limited [32]. Thus, it is desirable to sequester the phosphate from the wastewater for reuse instead of indiscriminate use of rock phosphate [32]. Phosphate accumulation is already reported in bacteria, but nitrate accumulation in bacteria is relatively rare. It is in the genus *Beggiatoa, Thioploca* and *Thiomargarita* that nitrate accumulation is observed in intracellular vacuoles [33–35]. Only recently nitrate accumulation from wastewater has been reported in the genus *Bacillus* [36]. Since nitrate and phosphate are both essentials for agriculture, but only a small fraction (12–30%) [7] of the applied nutrients is utilized by the plant, thus it becomes essential to trap these nutrients for reuse as well as environmental protection.

In order to address this upcoming environmental challenge, an alternative plant nutrient management strategy was developed with the following approach: (i) isolation and characterization of microbial consortium with ability to simultaneously accumulate nitrate and phosphate; (ii) utilize these microbes to prevent nutrient leaching from soil; and (iii) utilize these microbes with intracellular accumulated nutrients as biofertilizer.

2. Consortia development and characterization

Nitrate broth (Himedia M439) was used as the medium of choice for isolation of nitrate reducing microbial consortium. Two types of inoculum were used under both aerobic and anaerobic condition (in an atmosphere of carbon dioxide and nitrogen) at 37°C. The first type was the soil from East Calcutta Wetland (ECW) (22°27′ N, 88°27′E) which is known as the world's largest waste dumping ground and natural waste recycling center [37]. The reason for selecting soil from East Calcutta Wetland as the inoculum was that it was expected to harbor microbes with rich diversity as well as bioremedial ability. Since cultivation is the ongoing practice in this area, efficient strains with potential for promoting plant growth are expected to inhabit this area. The other inoculum was the biomass from a low-level radioactive waste treating microbial biofilm bioreactor removing mainly nitrate [38, 39]. This was expected to contain efficient nitrate reducers/accumulators due to its constant exposure to nitrate. Nitrate removal from the medium by the bacteria was set as the primary criteria for the selection of consortium. After 48 h of incubation, the nitrate concentration [40, 41] in the cell-free medium was checked. Of the four different combinations tested, two consortia were found to be efficient: anaerobic consortium from ECW (NB1) and aerobic consortium from bioreactor biomass

(BN7). They demonstrated 96 and 97.44% nitrate removal in 12 and 4 h by NB1 and BN7, respectively [39]. Another interesting feature of BN7 was its simultaneous accumulation of nitrate and phosphate from medium.

Both the cultures were also tested for phosphate removing ability as per standard procedure [30, 32] and demonstrated 23.88 and 48.2% removal with 565 and 1.14mg per gram wet weight of polyphosphate in NB1 and BN7, respectively. NB1 reduced 75–90% nitrate within a pH range of 5–12 with the maximum at pH 10 while that of BN7 was a range of 6–11 [39]. The optimum temperature range for NB1 was 30–40°C and that for BN7 was 25–37°C [39].

The effect of metals [viz., zinc ($ZnSO_4$), cobalt ($CoCl·6H_2O$), lead {$Pb(NO_3)$} and copper ($CuSO_4·5H_2O$)] on the nitrate reduction efficiency of NB1 and BN7 consortia was checked at two different concentrations, that is, 0.1 and 0.5 mM. It was compared to the reduction in the absence of metal salts (control) in both cases. The experiments were repeated thrice. The aerobic culture exhibiting growth along with nitrate reduction in the presence of different metal salts was checked for metal accumulation within the biomass using energy-dispersive X-ray fluorescence (EDXRF) analysis [39, 40]. While chromium (Cr), strontium (Sr) and cadmium (Cd) salts were inhibitory for the growth of the anaerobic consortium NB1 even at a concentration of 0.1 mM, the consortium showed growth in up to 0.5 mM concentration of copper (Cu), lead (Pb), cobalt (Co) and zinc (Zn). Being an anaerobic consortium, it was better preserved as glycerol stock while retaining its nitrate removal activity up to 24 days rather than stab or lyophilized culture as compared to BN7 [39].

16S rDNA based molecular characterization of both the consortia were done as per prior report [42]. The sequences obtained were subjected to NCBI nucleotide BLAST analysis, and novel sequences were submitted to GenBank. These sequences were then subjected to phylogenetic analysis using neighbor joining method. The rarefaction curves were drawn, and the richness (Shannon diversity index) and evenness (equitability index) of the population were determined as per standard procedure [37, 43, 44]. Mothur analysis was conducted using the data.

At the molecular level, NB1 was composed of novel organisms (GenBank JN626182-JN626198 and JN665074-JN665081) with closest identity in the ratio of 44:37:19 with *Pseudomonas* sp., *E. coli* and uncultured bacterium (**Figure 1a–c**) with poor diversity (Shannon diversity index 0.417) of evenly distributed population (equitability index 0.873). *Pseudomonas* sp. might be involved in nitrate removal as well as phosphate accumulation. BN7 on the other hand was composed of *Pseudomonas* sp.:*Azoarcus* sp.:uncultured bacterium: *Bacillus* sp. in the ratio of 20:31:46:3% in terms of 16S rDNA sequence similarity of its clones (GenBank GU644465 to GU644489). Like any enriched consortium in selective medium, BN7 reflected poor diversity (Shannon diversity index 0.39) of evenly distributed microbes (equitability index 0.83). Genus Pseudomonas and Bacillus were involved in phosphate accumulation and nitrate reduction [39].

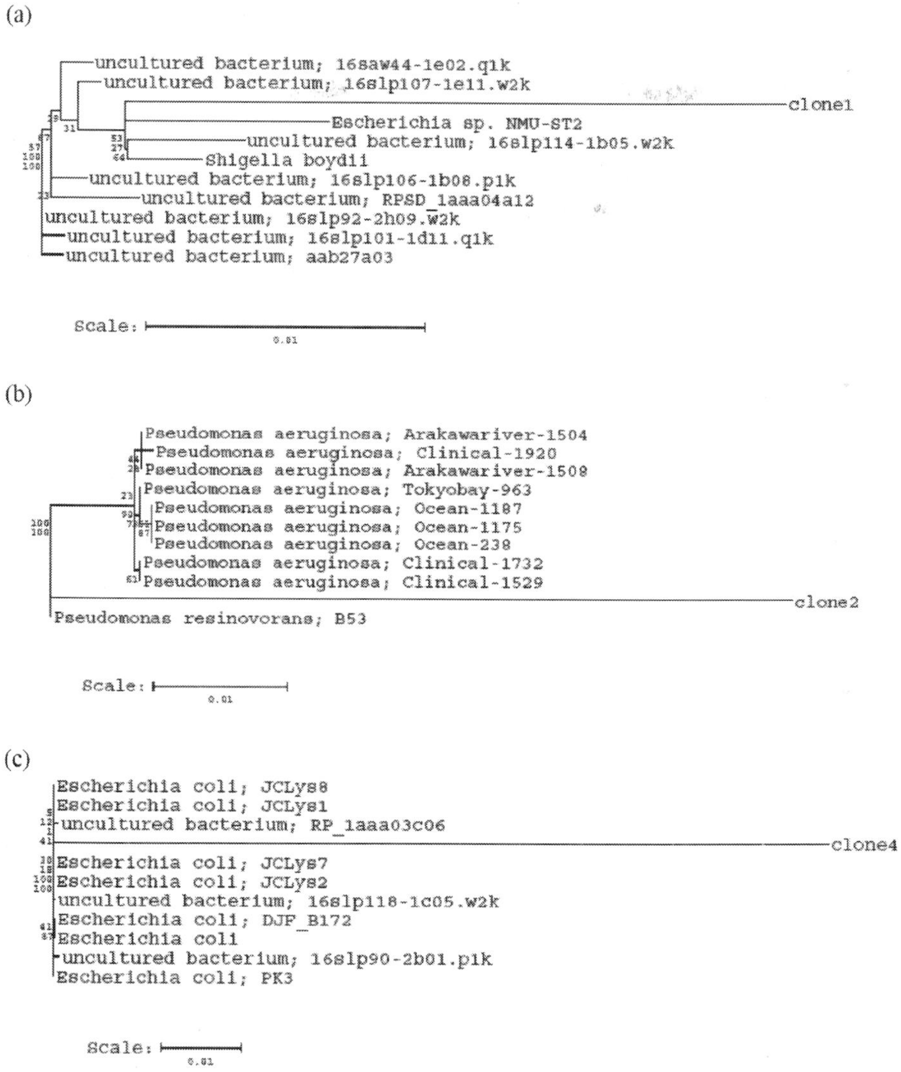

Figure 1. Phylogenetic trees constructed using neighbor joining method for the clones from the consortium NB1 showing maximum similarity with uncultured bacterium (a), Pseudomonas (b) and *E. coli* (c).

Mothur analysis revealed saturation of screening of the consortia which were different from each other (**Figure 2**; **Tables 1** and **2**).

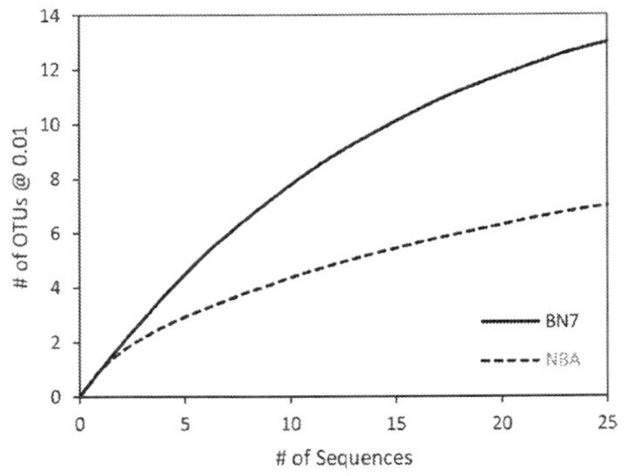

Figure 2. Rarefaction curve drawn for the consortium BN7 and NB1 reflecting saturation of screening for both the consortiums.

Comparison	dCXYScore	Significance
BN7-NB1	0.0206	<0.0001
NB1-BN7	0.0121	<0.0001

Table 1. Libshuff comparison showing that both libraries have a very different community structure.

Diversity index @ 0.01	BN7	NB1
N	25	25
S	13	7
Simpson (1/D)	18.75	3.03
95% LCI	12.90	1.96
95% HCI	34.32	6.69
Shannon (H)	2.47	1.41
95% LCI	2.22	0.99
95% HCI	2.72	1.82
H_{max}	2.84	1.67
Chao	15.00	8.00
95% LCI	13.29	7.09
95% HCI	26.96	17.68
Ace	16.25	10.08
95% LCI	14.49	7.45
95% HCI	20.07	28.24

Diversity index @ 0.01	BN7	NB1
Jackknife	18.00	10.00
95% LCI	11.80	5.20
95% HCI	24.20	14.80

Table 2. Diversity indices calculated for both the consortia.

3. Soil leaching

An experimental tub of dimension 18 cm × 12 cm × 17 cm (l × b × h respectively) (**Figure 3**), with surface area of 216 cm^2 and volume 3672 cm^3 filled up with 8.095 kg of soil, was set up for studying nitrate leaching in soil. In order to study the leaching process, outlets were made along the breadth of the tub at different heights of 3, 7, 11, 15 and 17 cm from the surface of the soil which facilitated in sample collection which in turn were assessed for the nitrate concentration [37, 38].

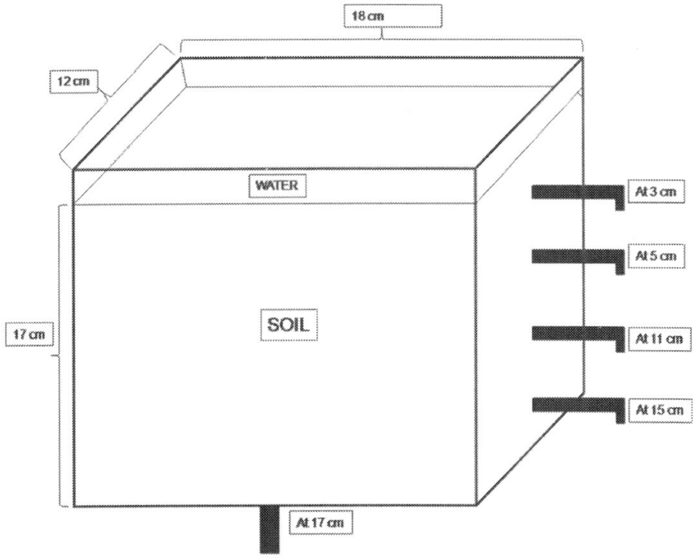

Figure 3. Schematic representation of the apparatus (soil filled tub) used for soil leaching experiment.

The experiment was carried out in four sets. For the first set (control), leaching of nitrate from soil in the presence of the native soil microbial population was tested. For this, water was poured into the soil filled tub. As the water seeped down, samples were collected from each outlet and analyzed for nitrate concentration [37, 38]. For the second and third set, the soil was inoculated with 100 ml of seed culture of BN7 and NB1, respectively. The system was left for 48 h for the consortium to colonize in the soil. Finally after 48 h, the leaching experiment was repeated as reported above to assess the nitrate released from the soil into the seepage water collected at different heights as a result of the interaction of soil native microbial population with the applied microbial consortia separately. For the fourth set, the combination of BN7 and NB1 in 1:1 ratio was applied and the experiment was repeated as in case of set two and three. The leaching of nitrate with and without external microbial consortium application was analyzed from the above experiments. This study was repeated thrice. In case of control, the soil interaction with the native microbial population as reflected through nitrate leaching was analyzed. In case of BN7 and NB1, these consortia were applied separately and the mixed impact of these consortia with the existing native soil microbial population was studied on the extent of nitrate leaching in water with traversed soil depth. In case of NB1 + BN7, the joint interaction of all the three consortium on nitrate leaching in soil was analyzed. From the results, it was observed that the application of the mixed formulation prevented leaching of nitrate from the soil resulting in decrease in the incidences of eutrophication due to soil nitrate leaching as documented in **Table 3**. It results in substantial reduction in nitrate leaching.

Level	Concentration of nitrate in seepage water at different levels in ppm							
	Distance from soil surface (cm)	Control	BN7	Difference in concentration (fold change)	NB1	Difference in concentration (fold change)	BN7 + NB1	Difference in concentration (fold change)
A	3	0	92.34	–	0	–	0	–
B	7	4.8	5.4	12.5	0	−100	0	−100
C	11	28.25	255.53	804.53	123.68	337.8	0	−100
D	15	75.1	425.7	466.84	154.82	106.15	4.36	−94.2
E	17	110.65	1160.27	948.59	120.6	8.99	12.83	−88.41
Correlation coefficient	–	0.94	0.82	–	0.88	–	0.79	–

Table 3. Tabular representation of the nitrate leaching from soil in the presence of different microbial consortia.

The correlation coefficient values indicate strong correlation between the depth of soil traversed by the applied water and the extent of nitrate leached in the presence of all the four treatments. Moreover, the prevention of leaching was complete at 11 cm of soil depth, indicating immobilization of nitrate in that zone. If this nitrate is made available to plants then this being the root zone for most of the plant, the productivity is expected to rise and the soil fertility is expected to be maintained. Also the phosphate accumulated inside as polyphosphate upon being released could be solubilized by the phosphatase released by the bacteria and made available to the plants. Both these phenomena are expected to strengthen the ability of this consortium (NB1 + BN7) to function as a biofertilizer. The nitrate and phosphate concentration in agricultural runoff could also be reduced by these microbes.

4. Plant growth promoting activity

Production of phytostimulator like ammonia, hydrogen cyanide (as plant protector), indole acetic acid, gibberellic acid (as plant hormones), phosphatase (to solubilize inorganic phosphate) and siderophore was tested for both the consortiums as per standard procedure [45]. NB1 produced 5.2 mg/100 ml and BN7 produced 1.64 mg/100 ml of ammonia with no hydrogen cyanide and siderophore production by either of them. Indole acetic acid (550 μg/ml) was produced by NB1 only. Both NB1 and BN7 produced enzyme phosphates, which were quantified to be 9.12 and 8.7 U/ml, respectively, with a final pH change to 4.11 and 6.3.

Since the consortium (NB1 + BN7) possessed plant growth promoting characters and also prevented leaching from soil, thereby making soil nutrients available to plants, both (NB1 and BN7) were tested for its effect on germination following soil application at the time of sowing, and the data were analyzed as per the standard protocol [45]. The data represent the combined effect of the native soil microbial population with the applied consortium. In order to assess the effect of only the combined consortia (NB1 + BN7) on germination in mung bean, the germination trial was repeated in germination tray using sterile soilrite mix kel006 (soil-free medium by Keltech Energies Limited, Bangaluru, India) and compared with that of control (uninoculated sterile soilrite). Application of either consortium improved the germination percentage, germination index and vigor index relative to the untreated control (**Table 4**).

Germination trial data	**Treatment set**		
Parameter	**Control**	**BN7**	**NB1**
Germination percentage	74.07 ± 22.45	98.15% ± 3.21	92.59 ± 8.49
Germination index	39.77 ± 9.39	75.95 ± 11.87	82.47 ± 11.23
Vigor index	1639.06 ± 366.67	1925.38 ± 490.02	1959.3 ± 632.25

Table 4. Represents data for germination trial with and without consortium application.

Even without any supporting microbes in the soil-free medium (Soilrite mix), this combination (NB1 + BN7) enhanced *Vigna radiata* (mung bean) germination (98%) as compared to the control (78%).

The consortia (NB1, BN7, NB1 + BN7) were further tested during pot trial (at Maulana Abul Kalam Azad University of Technology, India) and field trial for *Vigna radiata* var Samrat (developed by Indian Institute of Pulse Research, Kanpur, India) from Feb 2013 to May 2013 (spring/summer cultivation). The culture was applied only once at the time of sowing. For field trial, randomized block design with four replicates was carried out at Bidhan Chandra Krishi Viswavidyalaya Seed farm, Kalyani, Nadia, West Bengal, India as well as at State Department of Science and Technology facility, Salt Lake, Kolkata, West Bengal,

India. The sowing was done in the north south orientation in February 2013. The seeds post-germination were subjected to thinning on the 8th day post-sowing such that each 1 m^2 area contains a total of 40 plants (4 rows of 10 plants each). The inoculum applied on the day of sowing for field trial was 3.68 × 10^9 cells per plot (1 m × 1 m). The following parameters were monitored: plant height, number of branches, 50% flowering, 100% flowering, number of flowers, pod initiation, number of pods/plant, pod length, weight/pod, seeds/pod and weight of 100 seeds. In order to compare the data of the above-mentioned agronomic parameters as well as yield with that of conventional agriculture, simultaneously four (1 m × 1 m) plots were treated with chemical fertilizer. The chemical fertilizer (12.59 g) was applied in the ratio of N:P:K equals 20:40:40 (urea:single super phosphate:murated potash) for each 1 m × 1 m area. The total yield per hectare for each of the applications was monitored with respect to control (unfertilized). When applied together (NB1 + BN7) in field trials, the consortium significantly improved plant growth as compared to separate application (**Table 5**).

Parameters	Treatments				
	Control	NB1	BN7	NB1 + BN7	Chemical
Height of plants (cm)	37.86 ± 4.79	38.87 ± 10.27	40.25 ± 9	38.99 ± 6.79	31.34 ± 8.57
Number of branches	7.8 ± 0.63	7.9 ± 0.8	8.2 ± 1.3	8.9 ± 0.99	8 ± 1.41
Number of pods per plant	4.12 ± 3.09	10.25 ± 3.87	12.89 ± 4.98	11.85 ± 6.23	3.87 ± 2.69
Pod length (cm)	6.33 ± 0.86	7.65 ± 0.67	7.71 ± 1.31	8.07 ± 1.12	7.83 ± 1.05
Weight per pod (g)	0.41 ± 0.12	0.58 ± 0.23	0.53 ± 0.18	0.77 ± 0.22	0.53 ± 0.11
Seeds per pod	4 ± 1.58	4 ± 0.83	5 ± 1.15	7 ± 1.3	10 ± 0.83
Weight of 100 seeds (g)	3 ± 0.005	3.7 ± 0.45	3.59 ± 0.86	4.34 ± 0.46	4.27 ± 0.01

Table 5. Agronomic parameters for mung bean cultivation following chemical and biofertilizer application as compared to control (unfertilized) condition.

For every parameter, the combined application of NB1 + BN7 exhibited a better effect. Notably, the calculated yield per hectare was highest for NB1 + BN7 (2582.5 kg/ha) followed by chemical fertilizer (2017.5 kg/ha), BN7 (1802.5 kg/ha), NB1 (799.6 kg/ha) and the control (710.05 kg/ha). Thus, it offers potential advantage in meeting the increased food requirement in today's limited availability of land for agriculture. In addition, the consortia NB1 + BN7 also maintained soil fertility as revealed during the pot trial (**Table 6**).

In addition, each consortium (NB1, BN7, NB1 + BN7) could remove hydrocarbons such as metacil, pesticide and servo (lubricant) from the soil, suggesting that it has potential use in oil spill bioremediation.

Test parameters	Treatments				
	Unused soil	Control	NB1	BN7	NB1 + BN7
pH (1:2.5)	6.4	6.2	6.8	7.2	7.3
Conductivity (1:5) ds/m	0.091	0.086	0.108	0.13	0.079
Alkalinity (mg/kg)	225	187.5	225	225	187.5
Sodium (mg/kg)	156.67	150.16	138.25	119.05	168.65
Potassium (mg/kg)	69.9	60.25	44.46	54.43	76.11
Phosphate (mg/kg)	52.71	39.22	31.56	44.13	60.37
Amonical nitrogen (mg/kg)	87.5	73.5	89.25	70	99.75
Kjeldahal nitrogen (mg/kg)	96.25	82.25	85.75	78.75	108.5
Nitrate (mg/kg)	36.7	28	34.3	32.8	44.4
Nitrite (mg/kg)	27.2	20.8	25.4	24.3	32.9
Hydrocarbon (%)	0.136	0.041	0.004	0.004	0.09
Bulk density (g/cc)	1.11	1.05	1.12	1.16	1.11
Particle density (g/cc)	2.55	2.42	2.43	2.53	2.61
Pore space (%)	59.39	59.21	55.81	57.08	59.92
Water holding capacity (%)	53.25	56.13	50.4	50.52	52.94
Organic carbon (%)	1.36	1.23	0.95	0.82	1.91
Organic matter (%)	2.34	2.12	1.64	1.41	3.29
Available nitrogen (mg/kg)	113.75	105	117.25	99.75	138.25
Available potassium (mg/kg)	63.3	51.12	34.41	41.96	53.51
Available phosphorous (mg/kg)	17.2	12.8	10.3	14.4	19.7
Moisture (%)	2.91	2.7	1.89	1.65	2.76
Sand (%)	28.2	31.6	38.2	39.1	33.9
Silt (%)	43.4	42.5	36.8	37.5	37.5
Clay (%)	28.4	25.9	25	23.4	28.6
Textural classification	Clay loam	Loam	Loam	Loam	Loam

Source: Refs. [48–52].

Table 6. Soil nutritional quality analysis pre- and post-cultivation of mung bean during pot trial using standard methods..

5. Seed quality analysis

The seeds were lyophilized for 24 h and manually ground in the mortar and pestle; 0.2 g ground material was pelleted using Pelletizer (Technolab, Kbr Press) at 110 kg/cm^2. The mineral content of the pellets was assessed using energy-dispersive X-ray fluorescence (Jordan Valley EX–3600) analysis as per reported protocol [46, 47] at University Grant Commission-Department of Atomic Energy facility, Kolkata Center, India (**Table 7**).

Elements mg/kg (ppm)	Control	NB1	BN7	NB1 + BN7	Chemical	p-Value	Recommended by USDA
Zn	37.21 ± 2	44.57 ± 2.05	27 ± 3.02	29.06 ± 2.43	34.23 ± 2.58	0.04	26.8
Fe	68.34 ± 2.25	71.92 ± 1.66	68.45 ± 6.89	70.71 ± 0.57	67.21 ± 4.41	0.04	67.4
Mn	12.42 ± 0.44	12.74 ± 1.56	13.65 ± 1.43	15.46 ± 1.50	13.30 ± 0.64	0.02	10.35
Cu	13.30 ± 0.45	15.19 ± 0.56	15.66 ± 1.02	14.62 ± 1.39	14.49 ± 1.30	0.21	9.41
P	4242.09 ± 475.2	4604.71 ± 50.2	2429.97 ± 619.20	3741.01 ± 481.4947	1416.79 ± 574.18	0.003	3670.00
K	13,538.33 ± 491.76	13,830.88 ± 415.3	9651.83 ± 1546.293	11,807.17 ± 773.6117	10,943.22 ± 1349.72	0.18	12,460.00
S	2165.53 ± 288.35	2341.02 ± 63.25	1692.56 ± 199.5616	2037.44 ± 118.75	1575.90 ± 118.02	0.05	NA
Ca	2034.13 ± 149.41	2071.45 ± 214.95	1650.99 ± 410.549	1714.23 ± 79.81	1777.90 ± 396.11	0.04	1320.00

The commercially available fertilizer (Urea: Single Super Phosphate: Murated Potash) was applied in ratio of N:P:K equals 20:40:40 whereas in case of microbial biomass (N:P—2.52:1.51), 3.68 × 10^9 cells were added per plot (1 m × 1 m). The lyophilized seeds were manually grounded, and 0.25 g of the powder was converted into pellet and was analyzed by EDXRF for mineral content.

Table 7. Represents the elemental content of the seeds grown during control (unfertilized), chemical fertilizer as well as biofertilizer treatment.

The nutritional quality analysis like moisture [IS:4333(Part-II):2002], total protein (AOAC 920.87), available carbohydrate (AOAC 986.25), fat (AOAC 963.15), energy (Analytical Chemistry of Food by CS James:1995), ash content (AOAC 941.12), sugar (AOAC 923.09) and fiber (AOAC 985.29) content was carried out at SGS India Private Limited, Kolkata, India as per standard protocol (**Table 8**).

The statistical validation for the variation in elemental content of the seeds grown using varying treatments was carried out using single-factor ANOVA in Microsoft excel 2007. Here, the two hypotheses were as follows: null hypothesis H_0: no difference in elemental content with difference in treatment; alternative hypothesis H_1: significant difference in elemental content with difference in treatment. The level of significance was fixed at 5%. Based on a single-factor ANOVA, a significant variation was observed in the elemental content of the seeds produced after the treatments, especially in the Zn, Mn and Cu content between the control and NB1 + BN7 seeds. This clearly suggests that the consortium produces more elementally stable seeds. However, the overall nutritional quality of the seeds was maintained regardless of the treatment. The consortium exhibited similar trends for *Cicer arietinum* (chick pea) and *Abelmoschus esculentus* (ladies finger) cultivations.

Parameters	Treatment				
	Control	NB1	BN7	NB1 + BN7	Chemical
Energy value (kcal/100 g)	335.06	332.55	335.37	332	333.51
Total carbohydrate (g/100 g)	56.75	55.99	55.89	55.40	56.37
Protein (g/100 g)	23.61	23.46	23.19	22.86	23.79
Moisture (g/100 g)	14.85	15.87	16.19	16.82	15.46
Total ash (g/100 g)	3.86	3.73	3.64	3.87	3.98
Crude fat (g/100 g)	0.93	0.95	1.09	1.04	0.85
Total sugar (g/100 g)	3.20	3.13	2.95	3.07	3.20
Total dietary fiber (g/100 g)	15.65	15.38	15.18	14.99	15.18

Table 8. The nutritional quality of the seeds following cultivation under control (unfertilized), chemical fertilizer as well as consortium (NB1, BN7, NB1 + BN7) treatment.

6. Conclusion

The aim of this study was to develop an alternative strategy for plant nutrient management through microbial intervention. The objective of prevention of leaching of nitrate from soil was achieved through application of a 1:1 mixture of NB1 and BN7. It also ensured retention of nitrate within the root zone of soil. Being accumulators of nitrate and phosphate as well as producers of phytohormones with phosphatase activity, they could enhance germination while making the phosphate available for plant uptake. Thus, a single combination has the desired properties of a biofertilizer like phytohormone production, supplying of nutrients (nitrate and phosphate) resulting in higher yield of nutritionally enriched seeds. The unique selling points of this bioformulation are as follows: (i) its 21.88 times greater productivity (in case of mung bean) as compared to chemical fertilizer application and (ii) maintenance of soil fertility post-cultivation. Hereby, the remaining objections of multinutrient sequestration and reuse were effectively achieved. The wide range of pH and metal tolerance makes these consortia suitable for environmental application under varied conditions. These unique features of BN7 as well as NB1 + BN7 have been filed as Indian Patents 518/KOL/2011 dated April 11, 2011 and 203/KOL/2013 dated Feb 21, 2013. By this method, the nitrate concentration from agricultural runoff could be reduced substantially by using these microbes. All these properties point towards the future application of this innovation for bioremediation through nutrient sequestration from agricultural runoff as well as effluents and its reuse as biofertilizer with potential for environmental protection and agricultural sustenance.

Acknowledgements

Authors would like to acknowledge Mr. Vivek Singh for pot trial experiments using soil-free medium; Mr. Arpan Pal as well as Late Mr. Sourav Chakraborty for assistance during field trial experiments. The authors acknowledge Maulana Abul Kalam Azad University of Technology for the laboratory and computational facility. The authors would like to thank the following granting agencies for financial assistance: Ministry of Human Resource Development, Government of India (GOI) under the FAST scheme for providing the publication fee and fund for outsourcing services; Department of Biotechnology, GOI for M.Tech student fellowship; Indian Council for Agricultural Research, GOI under the NFBSFARA scheme for field trial related expenditure and seed quality analysis; Department of Atomic Energy under the BRNS scheme for consortium development and characterization as well as the World Bank under the TEQIP II scheme for student fellow.

Author details

Shaon Ray Chaudhuri[1,2]*, Madhusmita Mishra[3], Sonakshi De[3], Biswajit Samal[3], Amrita Saha[2], Srimoyee Banerjee[1,3], Abhinandan Chakraborty[3], Antara Chakraborty[3], Sonali Pardhiya[3], Deepak Gola[3], Joyeeta Chakraborty[3], Sourav Ghosh[1,3], Kamlesh Jangid[4], Indranil Mukherjee[2], Mathummal Sudarshan[5], Rajib Nath[6] and Ashoke Ranjan Thakur[1]

*Address all correspondence to: shaonraychaudhuri@tripurauniv.in

1 Department of Microbiology, Tripura University, Suryamaninagar, On lien from Maulana Abul Kalam Azad University of Technology, West Bengal (Formerly known as West Bengal University of Technology), India

2 Centre of Excellence in Environmental Technology and Management, Maulana Abul Kalam Azad University of Technology, West Bengal (Formerly known as West Bengal University of Technology), India

3 Department of Biotechnology, Maulana Abul Kalam Azad University of Technology, West Bengal (Formerly known as West Bengal University of Technology), India

4 Microbial Culture Collection, National Centre for Cell Science, Pune, India

5 Laboratory of Trace Elements, Inter University Consortium, Kolkata, Kolkata Centre, West Bengal, India

6 Faculty of Agronomy, Bidhan Chandra Krishi Viswavidyalaya, Mohanpur, Nadia, West Bengal, India

References

[1] Whitmore AP, Bradbury NJ, Johnson PA. Potential contribution of ploughed grassland to nitrate leaching. Agriculture, Ecosystems & Environment. 1992 Apr 30;39(3):221–33. doi:10.1016/0167-8809(92)90056-H

[2] Agrawal GD. Diffuse agricultural water pollution in India. Water Science and Technology. 1999 Feb 1;39(3):33–47 (Accessed 10th September 2016).

[3] Burkart MR, Kolpin DW, James DE. Assessing groundwater vulnerability to agrichemical contamination in the Midwest US. Water Science and Technology. 1999 Feb 1;39(3):103–12. doi:10.1016/S0273-1223(99)00042-6

[4] Giupponi C, Eiselt B, Ghetti PF. A multicriteria approach for mapping risks of agricultural pollution for water resources: the Venice Lagoon watershed case study. Journal of Environmental Management. 1999 Aug 31;56(4):259–69. doi:10.1006/jema.1999.0283

[5] Jørgensen LA. The cycling of nitrogen in the Danish agricultural sector and the loss to the environment. Water Science and Technology. 1999 Feb 1;39(3):15–23.

[6] Burkart MR, Stoner JD. Nitrate in aquifers beneath agricultural systems. Water Science and Technology. 2007 Jul 1;56(1):59–69. doi:10.2166/wst.2007.436

[7] Kinoshita T, Katoh T, Tsuji M, Kanada A, Inoue T. Downward movement of inorganic nitrogen in tea [*Camellia sinensis*] fields of yellow soil affected by different N-fertilizer application. Japanese Journal of Soil Science and Plant Nutrition (Japan). 2003. http://agris.fao.org/aos/records/JP2004002032 (Accessed 10th September 2016).

[8] Nakasone H, Yamamoto T. The impacts of the water quality of the inflow water from tea fields on irrigation reservoir ecosystems. Paddy and Water Environment. 2004 Aug 1;2(2):45–50. doi:10.1007/s10333-004-0039-2

[9] Wakida FT, Lerner DN. Non-agricultural sources of groundwater nitrate: a review and case study. Water Research. 2005 Jan 31;39(1):3–16. doi:10.1016/j.watres.2004.07.026

[10] Wakida FT, Lerner DN. Short communication potential nitrate leaching to groundwater from house building. Hydrological Process. 2006;20:2077–81. doi:10.1002/hyp.6143

[11] Cooper CM. Biological effects of agriculturally derived surface water pollutants on aquatic systems—a review. Journal of Environmental Quality. 1993;22(3):402–8. doi:10.2134/jeq1993

[12] Spalding RF, Exner ME. Occurrence of nitrate in groundwater—a review. Journal of Environmental Quality. 1993;22(3):392–402. doi:10.2134/jeq1993.00472425002200030002x

[13] Thorburn PJ, Biggs JS, Weier KL, Keating BA. Nitrate in groundwaters of intensive agricultural areas in coastal Northeastern Australia. Agriculture, Ecosystems & Environment. 2003 Jan 31;94(1):49–58. doi:10.1016/S0167–8809(02)00018–X

[14] Clifford D, Liu X. Ion exchange for nitrate removal. Journal (American Water Works Association). 1993 Apr 1:135–43 (Accessed 10th September 2016).

[15] Pinar G, Duque E, Haidour A, Oliva J, Sanchez-Barbero L, Calvo V, Ramos JL. Removal of high concentrations of nitrate from industrial wastewaters by bacteria. Applied and Environmental Microbiology. 1997 May 1;63(5):2071–3 (Accessed 10th September 2016).

[16] Awadallah RM, Soltan ME, Shabeb MS, Moalla SM. Bacterial removal of nitrate, nitrite and sulphate in wastewater. Water Research. 1998 Oct 31;32(10):3080–4.doi:10.1016/S0043-1354(98) 00069-4

[17] Eckford RE, Fedorak PM. Planktonic nitrate-reducing bacteria and sulfate-reducing bacteria in some western Canadian oil field waters. Journal of Industrial Microbiology and Biotechnology. 2002 Aug 1;29(2):83–92. doi:10.1038/sj.jim.7000274

[18] McAdam EJ, Judd SJ. Immersed membrane bioreactors for nitrate removal from drinking water: cost and feasibility. Desalination. 2008 Oct 31;231(1):52–60. doi:10.1016/j.desal.2007.11.038

[19] Choi JH, Maruthamuthu S, Lee HG, Ha TH, Bae JH. Nitrate removal by electro-bioremediation technology in Korean soil. Journal of Hazardous Materials. 2009 Sep 15;168(2):1208–16. doi:10.1016/j.jhazmat.2009.02.162

[20] Parvanova-Mancheva T, Beschkov V. Microbial denitrification by immobilized bacteria *Pseudomonas denitrificans* stimulated by constant electric field. Biochemical Engineering Journal. 2009 May 15;44(2):208–13. doi:10.1016/j.bej.2008.12.005

[21] Pawels R, Haridas A, Jose BT. Biological sulphate reduction with hydrogen in a jet loop biofilm reactor. International Journal of Scientific and Research Publications. 315. http://www.ijsrp.org/research-paper-0613/ijsrp-p1858.pdf (Accessed 10th September 2016).

[22] Hollo J, Czako L. Nitrate removal from drinking water in a fluidized-bed biological denitrification bioreactor. Acta Biotechnologica. 1987 Jan 1;7(5):417–23. doi:10.1002/abio.370070509

[23] Li H, Yang M, Zhang Y, Yu T, Kamagata Y. Nitrification performance and microbial community dynamics in a submerged membrane bioreactor with complete sludge retention. Journal of Biotechnology. 2006 May 3;123(1):60–70. doi:10.1016/j.jbiotec.2005.10.001

[24] Kesserű P, Kiss I, Bihari Z, Polyák B. Biological denitrification in a continuous-flow pilot bioreactor containing immobilized *Pseudomonas butanovora* cells. Bioresource Technology. 2003 Mar 31;87(1):75–80. doi:10.1016/S0960-8524(02)00209-2

[25] Van Rijn J, Tal Y, Schreier HJ. Denitrification in recirculating systems: theory and applications. Aquacultural Engineering. 2006 May 31;34(3):364–76.doi:10.1016/j.aquaeng.2005.04.004

[26] Cao GM, Zhao QX, Sun XB, Zhang T. Integrated nitrogen removal in a shell-and-tube co-immobilized cell bioreactor. Process Biochemistry. 2004 Jun 30;39(10):1269–73. doi:10.1016/S0032-9592(03)00256-5

[27] Konovalova V, Nigmatullin R, Dmytrenko G, Pobigay G. Spatial sequencing of microbial reduction of chromate and nitrate in membrane bioreactor. Bioprocess and Biosystems Engineering. 2008 Oct 1;31(6):647–53. doi:10.1007/s00449-008-0215-7

[28] Rathjen A, Hasch M. Reduction of nitrates by means of immobilized microorganisms. Engineering in Life Sciences. 2004 Oct 1;4(5):469–72. doi:10.1002/elsc.200403377

[29] Rezaee A, Godini H, Bakhtou H. Microbial cellulose as support material for the immobilization of denitrifying bacteria. Environmental Engineering and Management Journal. 2008 Sep 1;7(5):589–94 (Accessed 10th September 2016).

[30] Sayama M. Presence of nitrate-accumulating sulfur bacteria and their influence on nitrogen cycling in a shallow coastal marine sediment. Applied and Environmental Microbiology. 2001;8:3481–3487. doi:10.1128/AEM.67.8.3481.2001

[31] Krishnaswamy U, Muthusamy M, Perumalsamy L. Studies on the efficiency of the removal of phosphate using bacterial consortium for the biotreatment of phosphate wastewater. European Journal of Applied Sciences. 2009;1:06–15 (Accessed 10th September 2016).

[32] Van Kauwenbergh SJ, Stewart M, Mikkelsen R. World reserves of phosphate rock…a dynamic and unfolding story. Better Crops. 2013;97(3):18–20 (Accessed 10th September 2016).

[33] Otte S, Kuenen JG, Nielsen LP, Paerl HW, Zopfi J, Schulz HN, Teske A, Strotmann B, Gallardo VA, Jørgensen BB. Nitrogen, carbon, and sulfur metabolism in natural thioploca samples. Applied and Environmental Microbiology. 1999 Jul 1;65(7):3148–57. http://www.ncbi.nlm.nih.gov/pmc/articles/PMC91469/pdf/am003148.pdf (Accessed 10th September 2016).

[34] Schulz HN, Brinkhoff T, Ferdelman TG, Mariné MH, Teske A, Jørgensen BB. Dense populations of a giant sulfur bacterium in Namibian shelf sediments. Science. 1999 Apr 16;284(5413):493–5. doi:10.1126/science.284.5413.493

[35] Mußmann M, Hu FZ, Richter M, de Beer D, Preisler A, Jørgensen BB, Huntemann M, Glöckner FO, Amann R, Koopman WJ, Lasken RS. Insights into the genome of large sulfur bacteria revealed by analysis of single filaments. PLoS Biology. 2007 Aug 28;5(9):e230. doi:10.1371/journal.pbio.0050230

[36] DebRoy S, Bhattacharjee A, Thakur AR, RayChaudhuri S. Draft genome sequence of the nitrate-and phosphate-accumulating Bacillus sp. strain MCC0008. Genome Announcements. 2013 Feb 28;1(1):e00189–12. doi:10.1128/genomeA.00189-12

[37] Kundu N, Pal M, Saha S. East Kolkata Wetlands: a resource recovery system through productive activities. In: Proceedings of Taal 2007: The 12th World Lake Conference 2008 Jul 8 (vol. 868, p. 881).moef.nic.in/sites/default/files/nlcp/G%20-%20Pollution%20 Abatement/G-1.pdf (Accessed 10th September 2016).

[38] Mishra M, Jain S, Thakur AR, RayChaudhuri S. Microbial community in packed bed bioreactor involved in nitrate remediation from low level radioactive waste. Journal of Basic Microbiology. 2014 Mar 1;54(3):198–203. doi:10.1002/jobm.201200676.

[39] Ray Chaudhuri S, Sharmin J, Banerjee S, Jayakrishnan U, Saha A, Mishra M, Ghosh M, Mukherjee I, Banerjee A, Jangid K, Sudarshan M, Chakraborty A, Ghosh S, Nath R, Banerjee M, Singh SS, Saha AK, Thakur AR. Novel microbial system developed from low level radioactive waste treatment plant for environmental sustenance. In: Management of radioactive and hazardous waste. InTech; 2016. ISBN 978-953-51-4764-0 DOI:10.5772/63323.

[40] Cataldo DA, Maroon M, Schrader LE, Youngs VL. Rapid colorimetric determination of nitrate in plant tissue by nitration of salicylic acid 1. Communications in Soil Science & Plant Analysis. 1975 Jan 1;6(1):71–80. doi:10.1080/00103627509366547

[41] Adarsh VK, Mishra M, Chowdhury S, Sudarshan M, Thakur AR, Chaudhuri SR. Studies on metal microbe interaction of three bacterial isolates from east Calcutta wetland. doi:10.3844/ojbsci.2007.80.88

[42] RayChaudhuri S, Thakur AR. Microbial DNA extraction from sample of varied origin. Current Science. 2006 12: 1697–1700. http://www.researchgate.net/publication/237225066_Microbial_DNA_extraction_from_samples_of_varied_origin/file/3deec525bd625c8a61.pdf (Accessed 8th June 2014).

[43] Nasipuri P, Pandit G, Thakur AR, Chaudhuri SR. Microbial consortia from taptapani hot water springs for mining effluent treatment. American Journal of Microbiology. 2011;1(3):23–9.doi:10.3844/ajmsp.2010.23.29

[44] Nasipuri P, Pandit GG, Thakur AR, Chaudhuri SR. Comparative study of soluble sulfate reduction by bacterial consortia from varied regions of India. American Journal of Environmental Sciences. 2010;6(2):152–8. doi:10.3844/ajessp.2010.152.158

[45] Chakraborty U, Chakraborty BN, Chakraborty AP, Sunar K, Dey PL. Plant growth promoting rhizobacteria mediated improvement of health status of tea plants. Indian Journal of Biotechnology. 2013 Jan 1;12(1):20–31. doi:10.2135/cropsci1973.0011183X001300060013x. http://nopr.niscair.res.in/bitstream/123456789/16536/1/IJBT%2012%281%29%2020-31.pdf (Accessed 10th September 2016).

[46] Chowdhury S, Mishra M, Adarsh VK, Mukherjee A, Thakur AR, Chaudhuri SR. Novel metal accumulator and protease secretor microbes from East Calcutta Wetland. American Journal of Biochemistry Biotechnology. 2008;4(3):255–64.doi:10.3844/ajbbsp.2008.255.264

[47] Raychaudhuri S, Salodkar S, Sudarshan M, Thakur AR. Integrated resource recovery at East Calcutta Wetland: how safe is these? American Journal of Agricultural and Biological Science. 2007. doi:10.3844/ajabssp.2007.75.80

[48] Sacheti AK. Agricultural meteorology: instructional cum practical manual. New Delhi: National Council of Educational Research and Training; 1985, p. 53.

[49] Trivedi RK, Goel PK. Chemical and Biological methods for water pollution studies. Karad, India: Environmental Publication; 1984.

[50] Jackson ML. Soil chemical analysis. Englewood, N.J.: Prentice-Hall, Inc.; 1958, pp. 59–67.

[51] Black CA. Methods of soil analysis: chemical and microbiological properties, Part II. American Society of Agronomy; 1965.

[52] Katz M ed. Methods of air sampling and analysis. 2nd ed. Washington, DC: American Public Health Association; 1977.

Chapter 4

Mitigating Environmental Risks of Wastewater Reuse for Agriculture

Ahmed Al-Busaidi and Mushtaque Ahmed

Additional information is available at the end of the chapter

http://dx.doi.org/10.5772/65871

Abstract

The study was aimed to maximize and optimize treated wastewater reuse in conjunction with surface and ground waters resources. Moreover, environmental, agronomic and economic components were also considered. The project was funded by USAID and implemented in three countries (Oman, Tunisia and Jordan). In Oman, the study was done at Sultan Qaboos University experimental station field. Four types of waters (A: 50% of treated wastewater with 50% of groundwater, B: 100% of groundwater, C: 25% of groundwater with 75% of treated wastewater, and D: 100% of treated wastewater) were used to grow three different crops (okra, maize and sweet corn). Results showed no significant differences in soil physical and chemical properties with treatments irrigated with treated wastewater as compared to groundwater. On other hand, some chemical properties significantly increased ($p<0.05$) when treated wastewater was applied such as soil total carbon and some major elements (N, K, Mg). Crop physical analysis showed significant increases in plant productivity when plants were irrigated with treated wastewater and values of chemical properties were within the international standards. Crop biological analysis showed no effect on crop quality and all tested crops were free from any microbial contamination.

Keywords: treated wastewater, soil, fruits, yield, heavy metals

1. Introduction

Drought and overexploitation of conventional water resources present a critical problem in many regions of the world, especially the Middle East [1]. Therefore, water resources including nonconventional water should be well managed. Usage of treated wastewater (TWW) on agriculture can save fresh water resources and minimize the applications of chemical fertilizers. In many parts of the world, treated wastewater has been successfully used for irrigation, and many researchers have recognized its benefits [2, 3]. The continuous use of treated wastewater in irrigation increases the total soluble salts in the soil. The cation exchange capacity values are increased by increasing the period of using treated wastewater for irrigation, especially in the surface layer (0–30 cm). Moreover, Fe, Zn, Cu, Mn, Pb, and Co were increased by irrigation using treated wastewater as compared to virgin soil [4]. The use of treated wastewater for irrigation increased the mitotic index of divided cells, chromosome abnormality, and contents of individual amino acids. However, no differences in the profile of protein bands were observed between control and treated wastewater irrigation plants [5]. Whereas the accumulations of heavy metals in the edible part of some plants were detected which adversely affect human and animal health through the food chain [6].

Many countries have included wastewater reuse as an important component of water resources planning. Some countries like Oman have a national policy to reuse all treated wastewater effluents and have already made considerable progress toward this end [7]. Sultanate of Oman is one of the Middle East countries that is considered as the driest or semidriest region in the world with rapidly developing economy and a high population growth [8]. Soil and groundwater (GW) resources of good quality irrigation water have become limited. Rainfall is scanty to support crop production with annual mean rainfall of 100 mm. Therefore, its agriculture is almost fully dependent on groundwater [9]. Water resources augmentation together with conservation has been adopted by the government to combat the water shortage problem. The rapid development of Oman's urbanization, increase in population, and increase in agricultural production has led to high demand for water and urgent need to use treated wastewater as an alternative source of freshwater in agriculture. However, treated wastewater may contain high concentrations of salts, heavy metals, pathogens, and emerging pollutants with unknown effects on the ecological system [10]. High concentrations of heavy metals in plant fruits could affect human health and cause many environmental problems. However, the conjunctive use of treated wastewater and groundwater resources could be employed, helping to safeguard farmer's income and sustain food production. Despite this promising option, more research and education efforts are needed to ensure proper use of treated wastewater for agricultural production. Therefore, the study aimed to optimize treated wastewater reuse in conjunction with groundwater by taking into consideration their quantity and quality, in addition to the agronomic, environmental, and economic components.

2. Materials and methods

The field work was done in plots at the Agricultural Experiment Station, Sultan Qaboos University. Twelve plots (2.5 × 3.5 m each) were designed and sweet corn, okra, and maize crops were grown during the study. The plots were irrigated with four types of waters (A: 50% groundwater and 50% treated wastewater; B: 100% groundwater; C: 75% treated wastewater and 25% groundwater; and D: 100% treated wastewater) as shown in **Figure 1**. Plants were daily irrigated based on evapotranspiration (ETc). Soil samples were taken before and at the end of the study at a depth of 0–30 cm. Whereas plant samples were taken when the crop was mature and ready for analysis.

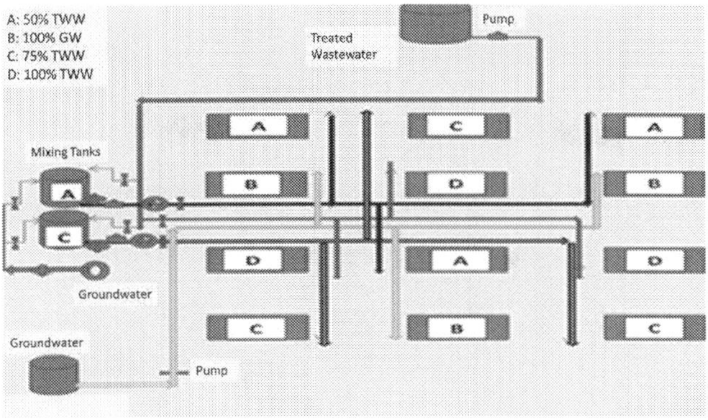

Figure 1. Field experimental design with different treatments.

Plants growth and yield of each crop irrigated by different waters were monitored. Fruits quality and quantity were assessed. Samples from soil and plants were taken for different physical, chemical, and biological analyses. All physicochemical analysis for soil and plants were done in soil and water labs (SQU) following standard methods [11] and using inductively coupled plasma (ICP) instrument for metals analysis. Soil and plant nitrogen (N) were analyzed in Rumais Research Laboratory (Ministry of Agriculture and Fisheries). Whereas biological analyses for crop samples were done in Muscat Municipality Laboratory.

The data were analyzed statistically using the analysis of variance (ANOVA) and the means were compared at the probability level of 5% using the least significant difference [12].

3. Result and discussion

3.1. Heavy metals in irrigation water

Growing conditions and the irrigation water are the most important parameters controlling plant life. **Table 1** demonstrates heavy metal concentrations in the irrigation waters that were used in the study. Comparing the used waters with national and international standards, it can be seen that elements concentrations mentioned in **Table 1** had lower values than applied standards. However, long-term application of some waters may accumulate some harmful elements in soil and plant tissues if mismanagement occurs. In some studies, it was found that wastewaters could carry appreciable amounts of trace toxic metals [13, 14] and concentrations of trace metals in sewage effluents vary from one city to another [15]. Although the concentration of heavy metals in sewage effluents are low, long-term use of these wastewaters on agricultural lands often results in the build-up of elevated levels of these metals in soils [15]. The results of Rattan et al. [15] reported high amount of Cr, Cu, Pb, Co, Ni, Mn, Cd, Fe, Zn, and As in sewage effluents compared to groundwater. Whereas soil organic matter was also increased in soil samples irrigated with sewage effluents compared to groundwater.

Water	Mn	Fe	Zn	Cu	Cr	Cd	Pb	Ni	B
Groundwater	0.002	0.013	0.013	0.008	<0.002	<0.001	<0.001	<0.001	0.295
Treated wastewater	0.002	0.016	0.064	0.024	<0.002	<0.001	0.066	<0.001	0.508
EPA Standard	0.200	5.000	5.000	0.500	0.100	0.010	0.100	0.100	0.750
FAO Standard	0.200	5.000	2.000	0.200	0.100	0.010	0.500	0.200	0.750
Omani Standard	0.500	5.000	5.000	1.000	0.050	0.010	0.200	0.100	0.750

*Summary of U.S. EPA guidelines for water reuse for irrigation [16].

Table 1. Comparing heavy metals concentration (mg/l) in the irrigation waters with national and international standards*.

3.2. Soil physicochemical properties

Quality of irrigation water could affect soil physicochemical properties. It could improve the soil quality by adding more nutrients or degrading the soil by adding toxic salts. Soil organic matter and total carbon are usually interconnected parameters and they are good indicators for soil fertility. In our study, some of them were found to be high in treated wastewater (TWW) compared to groundwater (GW) treatments. It is an expected result since treated wastewater is usually rich in nitrogen and other nutrients, which enrich soil and enhance plant growth (**Figure 2a** and **b**).

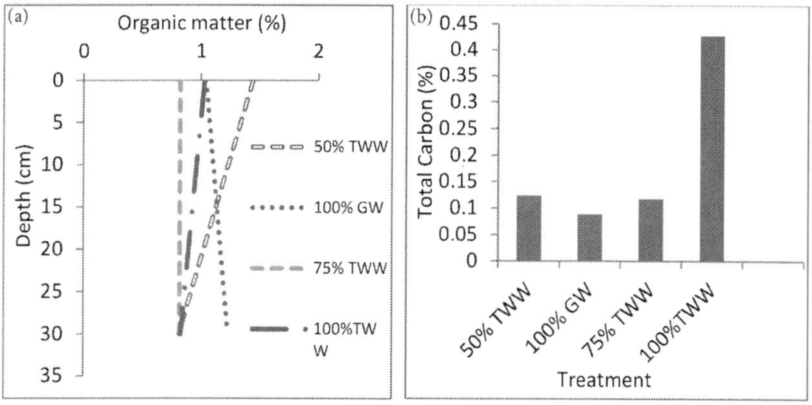

Figure 2. (a) Soil organic matter and (b) soil total carbon.

The presence of more nutrients (salts) in treated wastewater helps in keeping more water in plant root zone compared to groundwater (**Figure 3**). Nutrients as salts increased water viscosity and reduced evaporation process and as a result more water can be kept in the root zone [17].

http://dx.doi.org/10.5772/65871

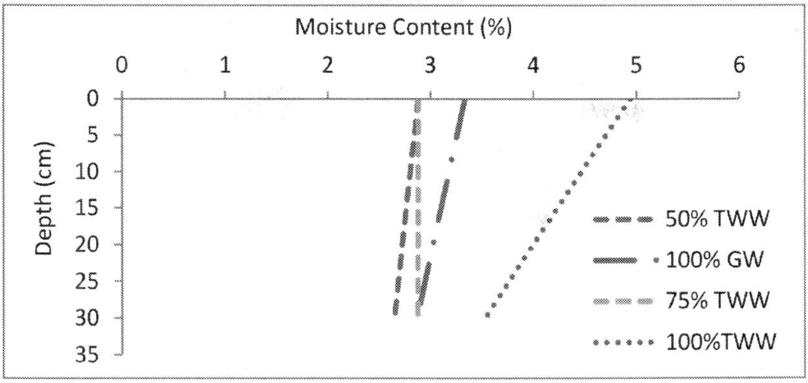

Figure 3. Soil moisture of all treatments.

Treated wastewater has a good amount of nutrients (salts). Therefore, it will add more salts to the irrigated soil and increase soil salinity compared to soil irrigated with freshwater (**Figure 4**). Salts are usually managed and reduced by leaching process.

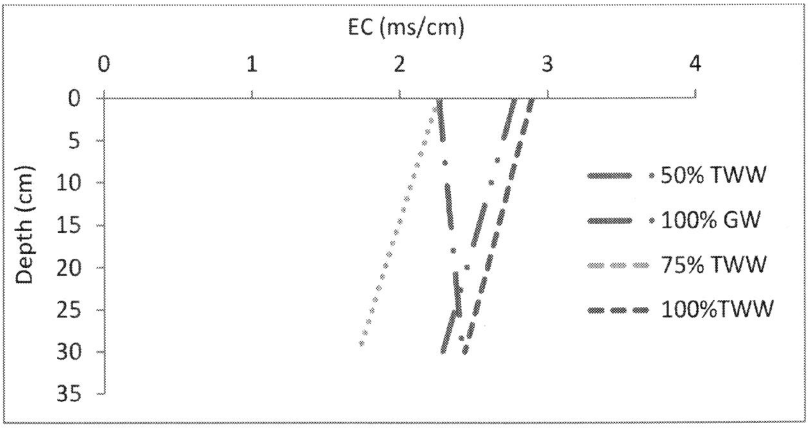

Figure 4. Soil electrical conductivity in all treatments.

In addition to organic matter, treated wastewaters have higher values for several nutrients compared to groundwater (**Figure 5**). These nutrients can improve soil fertility and later support plant growth. The variations in some elements' concentrations between treatments could be due to original nutrients concentrations in the soil and absorbance of those metals during plant growth. Mohammed and Mazahareh [10] found that treated wastewater irrigation increased soil salinity, soil phosphorous, potassium, iron, and manganese levels. They noticed that soil organic matter increased only in the topsoil.

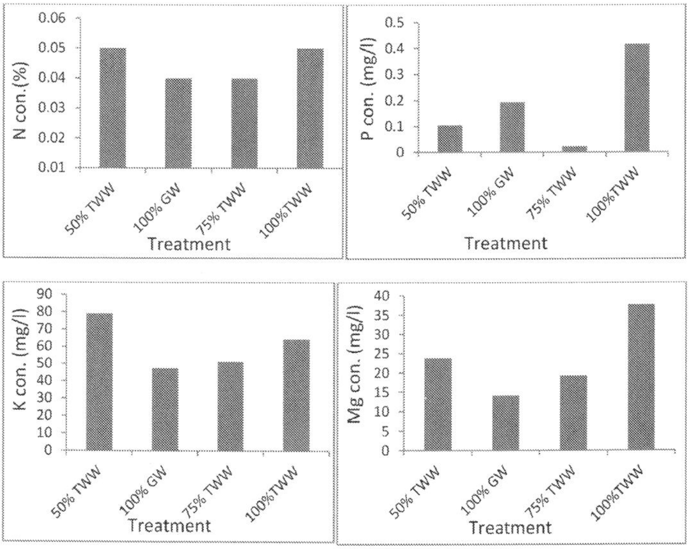

Figure 5. Mean concentration of some elements in soil samples.

Treatment	Mn	Cd	Fe	Zn	B	Ba	Cr	Co	Pb	Ni
50%TWW	0.018a	0.001a	0.330a	0.026a	0.166c	0.118a	0.043a	0.058a	0.196b	0.005a
100%GW	0.018a	0.001a	0.331a	0.001b	0.171b	0.123a	0.039a	0.060a	0.219a	0.011a
75%TWW	0.016a	0.001a	0.334a	0.001b	0.088d	0.087b	0.041a	0.061a	0.220a	0.005a
100%TWW	0.018a	0.001a	0.345a	0.003b	0.309a	0.110a	0.039a	0.062a	0.234a	0.001a

*Means in the column with same letter indicate no difference at Duncan's Multiple Range Test at $p < 0.05$.

Table 2. Mean concentration of heavy metals (mg/l) in soil samples*.

Checking soil for microelements (heavy metals) concentrations, it can be seen in **Table 2** that all values of heavy metals for both treatments (treated wastewater and groundwater) were very close to each other. However, some significant differences were found between some treatments which could be an indicator for long-term changes in soil chemical properties which is also found in Bansal et al. [18] and Palaniswami and Sree Ramulu [19] studies when they applied wastewater for long period. Rattan et al. [20] observed a build-up of Zn, Pb, Ni, Mn, Fe, Cu, Cr, Co, and As in the sewage-irrigated soils, over the well water-irrigated ones. Significant effect of irrigation through sewage water was observed in case of studied metals. There has been an enormous build-up in the available Fe content in the sewage-irrigated soils. Soils irrigated with groundwater and sewage water showed higher level of Cu and Zn. However, some sewage-irrigated soils accumulated more than 70 mg kg^{-1} total Zn, which could

http://dx.doi.org/10.5772/65871

cause a phytotoxicity problem [20]. Whereas Berry et al. [21] found that soil zinc and copper were not significantly affected by wastewater irrigation.

3.3. Crop physicochemical analysis

From **Table 3**, it can be seen that treated wastewater gave the best yield for all three crops compared to groundwater. The good supply of different nutrients from treated wastewater enhanced plant growth and improved plant productivity. Abohassan et al. [22] and Stewart et al. [23] have identified the beneficial effects of treated sewage water on some trees grown in Saudi Arabia and Australia. Shafiq et al. [24] found an increase of 24, 45, and 68% in maize total fresh biomass, dry yield, and grain yield irrigated by treated wastewater compared to groundwater. Same finding was also reported by Harati [25] in maize plants.

Treatment	Sweet corn	Okra	Maize
50% TWW	0.091	12.500	1.273
100% GW	0.141	11.091	1.193
75% TWW	0.085	10.556	1.160
100%TWW	0.090	13.958	1.593

Table 3. Average weight (kg) of some crops grown in the study.

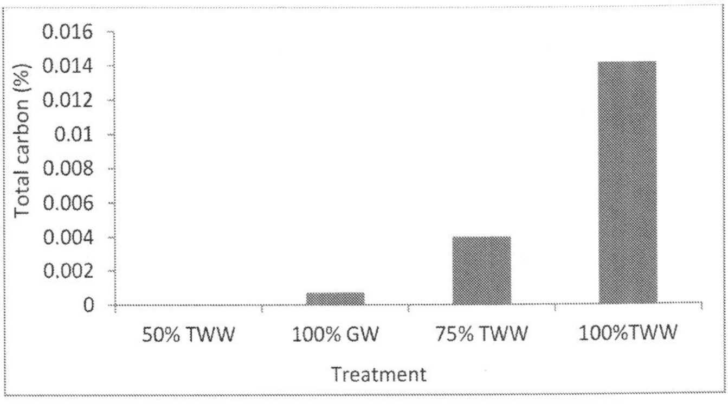

Figure 6. Percentage of total carbon in maize plant leaves.

Maize leaves were the best indicator for carbon content. Therefore, it can be seen from **Figure 6** that treated wastewater got the highest values compared to other treatments. It could be a reflection for what was found in water and soil samples. In a similar study done by Abd-Elfattah et al. [6], they found significant differences in metal content of plant leaves grown in soils irrigated with treated wastewater and plant leaves grown in soils irrigated with Nile water of both seasons.

Treatment	Element conc. (mg/l)								
Okra	Mn	Cd	Fe	Zn	B	Cr	Co	Pb	Ni
50%TWW	0.173d	0.001a	1.224c	0.357b	0.521c	0.060a	0.069a	0.252b	0.006b
100%GW	0.190c	0.001a	1.365b	0.364b	0.336d	0.071a	0.075a	0.229c	0.013b
75%TWW	0.242b	0.001a	2.372a	0.482a	0.745b	0.083a	0.087a	0.255b	0.127a
100%TWW	0.263a	0.001a	1.177d	0.495a	0.862a	0.057a	0.073a	0.300a	0.014b
Sweet corn	Mn	Cd	Fe	Zn	B	Cr	Co	Pb	Ni
50%TWW	0.177b	0.001a	1.295b	0.329c	0.073c	0.122b	0.091b	0.222c	0.068b
100%GW	0.204a	0.001a	1.582a	0.613a	0.492a	0.215a	0.100 a	0.191d	0.104a
75%TWW	0.152c	0.001a	1.584a	0.301c	0.122b	0.061c	0.070c	0.240b	0.011c
100%TWW	0.127d	0.001a	0.889c	0.400b	0.062c	0.064c	0.072c	0.444a	0.003d
Maize	Mn	Cd	Fe	Zn	B	Cr	Co	Pb	Ni
50%TWW	0.457b	0.001a	2.365b	0.256a	0.903a	0.136d	0.064a	0.210c	0.037a
100%GW	0.463a	0.001a	2.362b	0.219c	0.717c	0.146c	0.074a	0.213c	0.047a
75%TWW	0.366d	0.001a	2.279d	0.189d	0.454d	0.151b	0.075a	0.280a	0.040a
100%TWW	0.393c	0.001a	2.483a	0.226b	0.832b	0.181a	0.073a	0.241b	0.052a

*Means in the column with same letter indicate no difference at Duncan's Multiple Range Test at $p < 0.05$.

Table 4. Heavy metals concentration (mg/l) in tested crops*.

Standards/elements	Cd	Cu	Pb	Zn	As	Ni	Cr
WHO/FAO (2007)	0.2	40	5	60	–	–	–
European Union (EU 2006)	0.2	–	0.3	–	0.4	–	2.3
Indian Standard (Awashthi, 2000)	1.5	30	2.5	50	–	1.5	20

Source: CPCB [30].

Table 5. Guideline for safe limits of heavy metals in plants (mg/kg).

For soil, usually there is a direct relationship between salts found in the irrigation water and irrigated land. Whereas, for plants, root selectivity and present of salts in different forms could play a role in elements movement and translocation from soil to plant. From **Table 4**, it can be seen that concentrations of many elements were significantly ($p < 0.05$) different from one treatment to other. However, microelements in the edible parts of all crops grown in the field were not that high and they were within the international standards (**Table 5**). Same results were reported by Abdelrahman et al. [26] when they observed no significant difference between fresh and treated wastewater with regards to heavy metals accumulation in grown crops. Moreover, this finding was supported by Pescod [27] study, when he concluded that the concentrations of heavy metals in seeds were within normal level when treated wastewater effluent was used. Such results make it clear that heavy metal in soil are not readily bio-available for crop uptake and do not represent a threat to quality of crop consumption.

In general, treated wastewater contains variable amounts of nutrient elements and heavy metals. Availability and translocation of these elements to and within the plant tissues is highly dependent on the environmental conditions as well as their concentration and ratios in the plant organs [28]. Same results were also found by Mahdi et al. [29] when they reported that concentration of nutrient elements of different crops indicated that the crop nutrient uptake is affected by tree age and species. Longer exposure to treated wastewater did not indicate major effects on fruit minerals, including heavy metals. Sampling over longer period of time is needed to confirm the changes in nutrient composition over time. Therefore, in the present study it can be seen that treated wastewater treatment sometime got the highest values for heavy metals which could be an indication for heavy metal accumulation with long-term application if treated wastewater is used without proper management. This prediction could be similar to Abd-Elfattah et al. [6] findings when they found a significant difference in fruit contents of heavy metals and trace elements (Pb, Cd, Ni, Cu, Mn, Fe, Zn) between fruits produced by treated wastewater compared with Nile water in both seasons. The accumulation of heavy metals in the edible part of plant was detected which adversely affects human and animal health through the food chain [6].

Finally, the findings of this study are supported by many researches. As such, Omran et al. [31] found no significant problems with orange trees when they were irrigated with treated sewage water. Furthermore, in Hamad et al. [32] study, toxicity problems for some metals (Cd, Hg, Cr, Pb) in tested crops due to irrigation with treated wastewater was not observed. In the Sultanate of Oman it was found that treated sewage water did not cause any phyto-toxicity symptoms in date palm leaves and fruits [33]. Therefore, it can be concluded that proper management of wastewater irrigation and periodic monitoring of soil and plant quality parameters are required to ensure successful safe long-term wastewater irrigation [34].

3.4. Crop biological analysis

Usually microbial analyses are the direct indicators for microbial contaminations in different crops. In this study, the edible part of grown crops was checked by Muscat Municipality laboratory and different microbes were analyzed such as coliform bacteria, *Escherichia coli*, and *Salmonella* spp. All tested samples were free from any microbial contamination. This finding was supported by Mexican and Tunisian studies where sewage effluent at different levels of treatment has been employed to irrigate various crops. It has been used with no serious effect on man and plants [35].

4. Conclusion

The use of treated wastewater for irrigation is increasingly being considered as a technical solution to save fresh groundwater, minimize soil degradation, and improve soil fertility. In this study, usage of treated wastewater irrigation as compared to groundwater did not affect significantly some soil physical and chemical properties. Whereas some chemical properties such as major elements (N, K, Mg) and total carbon were significantly increased when treated wastewater was applied. Concentrations of heavy metals were increased in soils irrigated with treated wastewater compared to groundwater. The differences in heavy metals concentrations of all treatments were small and data of all treatments was close to each other.

Treated wastewater is a rich source of nutrients and provides most nutrients that are necessary for crop growth. Therefore, treated wastewater improved significantly plant productivity compared to groundwater treatments. Whereas small increase was noticed with some chemical properties of plants irrigated with treated wastewater compared to groundwater. However, all measured values were within the international standards. Biologically, all tested crops were free from any microbial contaminations. In general, most crops gave higher yield with wastewater irrigation and reduced the need for chemical fertilizers, resulting in net cost savings to farmers. Therefore, it can be concluded that treated wastewater is an important source of water for agricultural production and to avoid any health or environmental problems, quality of treated wastewater should be monitored with time.

Acknowledgements

The authors like to thank the staff from Sultan Qaboos University and Ministry of Agriculture and Fisheries for their support in collecting and analyzing soil, water, and plant samples. Special thanks to USAID (FABRI) for their financial support.

Author details

Ahmed Al-Busaidi* and Mushtaque Ahmed

*Address all correspondence to: ahmed99@squ.edu.om

Department of Soils, Water and Agricultural Engineering, College of Agricultural & Marine Sciences, Sultan Qaboos University, Muscat, Oman

References

[1] World Bank: Renewable Energy Desalination: An Emerging Solution to Close the Water Gap in the Middle East and North Africa. Washington, DC: World Bank. 2012.

[2] Mujeriego R: Salal Golf course irrigation with reclaimed waste water. Sci. Technol. 1991; 24 (9): 161–172.

[3] Levine A, Sanot A: Recovering sustainable water from waste water. Environ sci. Technol. 2004; 38 (11): 201A–208A.

[4] Selem MM, Amir S, Abdel-Aziz SM, Kandil MF, Mansour SF: Effect of irrigation with treated waste water on some chemical characteristics of soils and plants. Egypt. J. Soil Sci. 2000; 40 (1/2): 49–59.

[5] Hossni YA, El-Tarras A: Safety of treated waste water in the irrigation of plants. Egypt. J. Hortic. 1997; 24 (2): 261–270.

[6] Abd-Elfattah A, Shehata SM, Talab AS: Evaluation of irrigation with either raw municipal treated waste or river water on elements up take and yield of lettuce and potato plants. Egyptian J. Soil Sci. 2002; 42 (4): 705–714.

[7] Al-Ajmy I: Wastewater in the Sultanate and its effects on environment. In International conference on wastewater management and its effect on the environment in hot and arid countries, Ministry of Regional Municipalities, Environment and Water Resources, Sultanate of Oman. 2002.

[8] MAF (Ministry of Agriculture & Fisheries): General soil of the Sultanate of Oman, Directorate General of Agriculture, Ministry of Agriculture and Fisheries. Sultanate of Oman. 1990. p. 50.

[9] MRMWR (Ministry of Regional Municipalities and Water Resources) annual report: Environment in the Sultanate of Oman. Muscat, Oman. 2004; Vol. XIV, Issue No. 53.

[10] Mohammed MJ, Mazahareh N: Changes in soil fertility parameters in response to irrigation of forage crops with secondary treated waste water. Commun. Soil Sci. Plant Anal. 2003; 34: 1281–1294.

[11] Gee GW, Bauder JW: Particle-size analysis. In: Klute, A. (ed), Methods of Soil analysis, Physical and Mineralogical Methods, Part 1. 2nd ed. Agronomy series No. 9. American Society of Agronomy and Soil Science Society of America Inc. Publ., Madison, Wisconsin, USA. 1986; pp. 383–404.

[12] SPSS Inc: SPSS Base 8.0 for Windows User's Guide. SPSS Inc., Chicago. 1998.

[13] Brar MS, Mahli SS, Singh AP, Arora CL, Gill KS: Sewer water irrigation effects on some potentially toxic trace elements in soil and potato plants in northwestern India. Can. J. Soil Sci. 2000; 80: 465–471.

[14] Yadar RK, Goyal R, Sharma RK, Dubey SK, Minchas RS: Post irrigation impact of domestic sewage effluent on composition of soils, crops and ground water—a case study. J. Environ. Int. 2002; 28 (6): 481–486.

[15] Rattan RK, Datta SP, Chandra S, Sahran N: Heavy metals and environmental quality: Indian scenario. Fertile. News. 2002; 47 (11): 21–40.

[16] U.S Environmental Protection Agency: Guidelines for Waste Water Reuse, Municipal support office of wastewater management. Washington, DC, EPA/625/R-04/108. 2004.

[17] Al-Busaidi A, Cookson P: Leaching potential of sea water. J. Sci. Res. 2005; 9: 27–30.

[18] Bansal RL, Nayyar VK, Takkar PN: Accumulation and bioavailability of Zn, Cu, Mn and Fe in soils polluted with industrial waste water. J. Indian Soc. Soil Sci. 1992; 39: 795–799.

[19] Palaniswami C, Sree Ramulu US: Effects of continuous irrigation with paper factory effluent on soil properties. J. Indian. Soc. Soil Sci. 1994; 42: 139–140.

[20] Rattan RK, Datta SP, Chhankar PK, Suribabu K, Singh AK: Long term impact of irrigation with sewage effluents on heavy metal contents in soils, crops and groundwater—a case study. Agric. Ecosyst. Environ. 2004; 109: 310–322.

[21] Berry WL, Wallace A, Lunt OR: Ultilization of municipal wastewater for culture of horticultural crops. Hort. Sci. 1980; 15: 169–171.

[22] Abohassan AA, Kherallah IE, Kandeel SA: Effect of sewage effluent irrigation regimes on wood quality of *Presopi juliflera* grown in Riyadh Region. Arab Gulf J. Sci. Res. 1988; 136 (1): 45–53.

[23] Stewart HTL, Hopmans P, Flinn DW: Nutrient accumulation in trees and soil following irrigation with municipal effluent. Australian Environ. Pollut. 1990; 63(2): 155–177.

[24] Shafiq M, Hassan I, Hassan Z: Influence of irrigation methods on the productivity of summer maize under saline/sodic environment. Asian Journal of Plant Sciences. 2002; 1 (6): 678–680.

[25] Harati M: Study on heavy metal accumulation in different parts of corn irrigated by sewage in south of Tehran. MSc Thesis. Tehran University. 2003.

[26] Abdelrahman HA, Alkhamisi SA, Ahmed M, Ali H: Effects of Treated Wastewater Irrigation on Element Concentrations in Soil and Maize Plants. Conference Proceeding. Sultan Qaboos University, Oman. 2011.

[27] Pescod M: Wastewater treatment and use in agriculture. Bull. FAO #47 (125) (Rome). 1992.

[28] Izzo K, Scagnozzi A, Belligno A, Navari-Izzo F: Influence of NaCl treatment or Ca, K, Na interrelations in maize shoots. In: M.A.C. Frageso and M.L. Beusichem (Eds.), Optimization of Plant Nutrition. Kluwer Academic Publishers, Dordrecht, The Netherlands. 1993, pp. 577–582.

[29] Mahdi OE, Salama SB, Consolacion EC, Al-Solomi M: Effect of treated sewage water on the concentration of certain nutrient elements in date palm leaves and fruits. Commun. Soil Sci. Plant Anal. 1998; 29: 5–6.

[30] CPCB (Central Pollution Control Board): Parivesh, Newsletter from CPCB. 2002. Available at /http://www.cpcb.delhi.nic.in/legislation/ch15dec02a.htms

[31] Omran MS, Waly TM, Abo-Eleinain EM, Elnashar BMB: Effect of sewage irrigation on yield, tree components, and heavy metals accumulation in navel orange trees. Biol. Wastes BIWAED. 1988; 23: 17–24.

[32] Hamad I, Nizam AA, Suleiman MS: Biological detection of contamination in vegetables irrigated with brackish and underground water. In: Proceedings of the Arab Regional Workshop on the Use of Saline, Brackish, and Treated Sewage Effluent Water in Irrigated Agriculture, Muscat, Sultanate of Oman. 1995. pp. 481–531.

[33] El Mardi MO, Salama SB, Consolacion EC, Al-Shabibi MS: Effect of treated sewage water on vegetative and reproductive growth of date palm. Commun. Soil Sci. Plant Anal. 1995; 26 (11&12): 1895–1904.

[34] Rusan M, Hinnawi S, and Rousan L: Long term effect soil and plant quality parameters. Desalination. 2007; 215: 143–152.

[35] Strauss M: Sewage Treatment for Wastewater Reuse. Wastewater Reuse Seminar, Institute of Health Sciences, Wattyah, Muscat, Sultanate of Oman. 1987.

Section 2

Hazards and Treatment of Organic Compounds in Wastewater

Chapter 5

Spreading of Antibiotic Resistance with Wastewater

Sadik Dincer and Esra Sunduz Yigittekin

Additional information is available at the end of the chapter

http://dx.doi.org/10.5772/66188

Abstract

The recent statistics show that the world's population is rapidly increasing. This increase negatively affects the water resources and it increases the water demand progressively. Along with the increase in the world's population, the insensible use of water resources, pollution, and drought lead to the increasing reduction of water resources. Due to these factors, all countries, primarily developed countries, have started looking for new water resources. This search has been extended to extraterrestrial water. However, the existing technology and opportunities direct countries toward the purification of wastewater rather than searching for new water resources. For the reasons outlined above, purification and recycling of wastewater become important. In addition to the natural resistance of microorganisms against antibiotics, a resistance also arises because of the unconscious and overuse of antibiotics. This resistance spreads through wastewater progressively. Antibiotic resistance shows an increase according to the scientific data. In order to prevent the resistance, it is of capital importance to treat the wastewater in which the domestic pollution burden is high. In this study, the role of domestic wastewater in the occurrence and spread of antibiotic resistance will be revealed.

Keywords: antibiotics, spreading antibiotic resistance, water, wastewater, domestic wastewater

1. Introduction

Water contains millions of microscopic living beings within itself. The plenty amount of water is accessible on our planet for living beings to maintain their vital activities.

Along with 14 billion m^3 of water, 97.5% of it is salty water, 2.6% of it is freshwater, and 0.8% of the total amount of water is present as freshwater in the state of constant vaporization, precipitation, and flow. Water scarcity is indicated as one of the main problems of the twenty-first century in the whole world, and for this reason, lives of many people depend on the right usage of water. People need water primarily for civic, industrial, and agricultural areas. However, water is regarded as a limited source. For the fact that water resources become insufficient and decrease in quality creates serious concerns. The population increase, urbanization, agricultural practices, and industrialization increase the water demand. Wastewater treatment is built for the purpose of reducing the pollution by removing pathogens, nutrients, and biodegradable substances, and protecting public health and the environment. Furthermore, with the increase in water demand, the recycling of wastewater has been brought into question [1].

Wastewaters are divided into two groups such as domestic and industrial. Domestic wastewater can originate from house, workplace, and hospital because of its content. The complete treatment of this type of wastewater is impossible even if it goes through many stages. This situation causes many problems. The emergence and spread of antibiotic-resistant bacteria are the leading reasons for this problem and they make humans and animals sick.

According to American Centers for Disease Control and Prevention (CDC), antibiotic resistance constitutes one of the most important health problems of a country. In America, it is estimated that 2 million people become sick and 700,000 people die worldwide because of resistant bacteria. In a report of 2013, CDC indicated that the usage of antibiotics in the production of food animals causes the emergence of resistant *Campylobacter* that is contagious to humans. Resistance genes can be transferred between zoonotic bacteria types, among the bacteria species, through food chain and contact with feces of ill animals and contaminated environment [2].

Antimicrobial resistance causes many problems in humans and animals in the case of the spread through wastewater, spread wastewater treatment output's being low, and the usage of these waters in agricultural practices and irrigation fields, the emergence of many antibiotic-resistant microorganisms.

2. Significance of water

Water is an essential substance which is necessary for vital activities such as nourishment, circulation, respiration, excretion, and reproduction to occur in every period of human life. At the same time, water itself is a habitat as one of the basic elements in nature while forming a habitat. The presence of water in a habitat and its quality are extremely important for life [3].

As the most important one of all natural sources for all living beings, water is a habitat and it contains millions of microscopic living beings. It constitutes approximately three-quarters of the Earth [4].

Water is crucial for the life of living beings as the most common natural resource on the Earth. Seventy-five percent of the Earth's surface, seventy percent of the human body, and seventy-eight percent of the blood consist of water [5].

Ninety-seven percent of the water body on the Earth consists of oceans and seas, two percent of lakes, rivers, and underground waters, and one percent of glaciers and snows. Water has been used during the development of civilizations for many purposes such as personal hygiene, agricultural irrigation, industrial production, and electric power production [6].

Water submerges more than 70% of the Earth as fresh and salty water and these environments are defined as aquatic environments [7]. Salty water constitutes more than 96% of the water on the Earth. More than 68% of the present freshwater is found in ice and glaciers. In this way, it is considered that water stays in stock. Thirty percent of freshwater consists of underground water. Two-thirds of underground waters are located deeper than 800 m. Surface freshwater sources such as rivers and lakes constitute 93,100 km^3 (22,300 cubic miles), which is 1/700 of 1% of all water on the Earth [8].

The minerals, salts, and sulfates contained in the water are very important along with its other characteristics. The presence of these substances in a certain amount in water is essential for life while their presence in small or greater quantities continually affects life in a negative way. At the same time, water is a habitat. The pollution of this environment creates danger for life [9, 10].

An increase in industrialization, urbanization, and population that started at the beginning of the twentieth century, and an increase in the use of natural resources have caused the emergence of the problems called the environmental pollution jeopardizing human life. An increase in the variety and amount of solid and liquid wastes that are disposed into the environment causes air and water pollution [6].

Water is a component which is significant for the life cycle of all living beings on the Earth. While three-quarters of the Earth is covered by water, two-thirds of the human body is covered by water. This rate significantly affects all living beings while being important for both the Earth and human beings. Water has many important roles from systems in the living organism to cellular functions. Even a small decrease in the amount of water can endanger life.

Since the existence of the Earth, all civilizations have settled in the places around or close to water. This shows us that water is a functional substance which is completely life-oriented. The decrease in the amount of water arises from both the environmental pollution and unconscious consumption. Because of this decrease, countries are in search of new sources and wars break out.

3. Water consumption

Population increase on the Earth leads to the decrease of water and pollution of clean and potable water. This will cause water scarcity in the future. Rivers and lakes constitute most of the water that people use daily. The pollution of these water sources will create water shortage. The amount of water that meets our needs is 0.25% of all water sources on the Earth [11].

About 97.39% of 1384.10^9 km^3 water on the Earth is found in the oceans and seas. The remaining 2.01% consist of glaciers and 0.60% consists of underground water, lakes, and rivers. This situation shows that the available freshwater supply constitutes quite a small amount of all water sources on the Earth [12].

The world's population that is approximately 6 billion is able to use 54% of the renewable surface and underground water supplies. It is considered that this rate will increase to 70% with the population's increase as the conditions of use remain the same. At the same time, it is estimated that 90% of the present freshwater sources will be used with the increase of life standards and the increase of water usage per person. For other living beings, there will be 10% of the available water supply. It is indicated that there will not be enough water for environmental and ecological functions because of the population increase and unconscious use of water resources [13].

Water resources are also used in a sectorial aspect besides meeting daily needs. The usage of water is classified as agricultural, industrial, and domestic sectors [14].

Sixty-nine percent of the freshwater resources on the Earth are used for agriculture, twenty-three percent are used for industry, and eight percent are used for domestic purposes. These rates differ from continent to continent. For instance, while the rates of agricultural, industrial, and domestic usage of water in Africa are 88, 5, and 7%, respectively, these rates in Europe are 33, 54 and 13%, respectively [15].

Water consumption in the world has increased 10-fold since 1900. In the studies conducted, it is determined that water consumption will increase 17% in agriculture, 20% in industry, and 70% in domestic consumption in 2015. Moreover, it is told that 20% of 6 billion world population is deprived of clean water resources. The water amount per person decreased to 7300 m^3 in 2000 while it was 16,800 m^3 in 1950 [16–18]

While it is estimated that world population will be 8 billion, it is considered that water consumption per person will decrease to 4800 m^3 in 2025. This decrease in consumption will arise from water resources shortage. Furthermore, the present available water resources will be polluted in 2025 and then water will not be provided [19]. It is estimated that the curve of the increase in water demand and the curve of the decrease of clean water resources will intersect in 2023.

Recent studies show that population growth will increase the consumption of water. However, this situation is inversely proportional to the number of water resources. Due to the decrease of clean and available water resources, the quantity of water per person has decreased. The reason for these situations is water scarcity which arises from the unconscious use and pollution.

4. Water pollution

Water pollution has a negative effect on public health and ecology because of the degradation of water quality and natural balance. Water pollutants contain surplus metal, some radioactive isotopes, nitrogen, phosphorus, sodium, and other beneficial and necessary elements along with especially some faecal originated pathogenic bacteria, parasites, and viruses which can be human or animal originated.

Mixing of any organic, inorganic, radioactive, or biologic substance that inhibits or disturbs the usage of water resources by impairing their quality into water is called water pollution [20, 21].

The reasons of water pollution are particularly domestic, industrial, agricultural, physical, chemical, radioactive, and microbial pollution.

Domestic wastewater contains organic and inorganic substances that are suspended, colloidal, and dissolved. Domestic wastewater consists of organic foods such as too much carbon, nitrogen, and phosphorus and highly concentrated microorganisms [22]. With the increase in urbanization, the flow of domestic waste into water through sewerage system also increased. In particular, detergents which are used in washing machines, oils poured out into lavabo, and the dispersion of wastes that should be accumulated in dustbins and recycled into the environment cause water pollution [23]. The characteristics of industrial wastewater differ from industry to industry [22].

Apart from domestic and industrial wastewaters that are discharged into water sources without being treated, the unconscious fertilization and unconscious usage of agricultural pesticides are also the reasons for pollution. These pollutants become crucial with negative effects on the water resources regarded as inadequate according to the world average, environmental, and public health and in terms of economy [6]. In the fields close to water, the incorrect ploughing mixes into water through the wind and causes pollution in water [3]. An increase in the usage of synthetic manure and pesticides in agriculture and industry and chemical substances that are used in the industry create a risk of water pollution [23].

Industrial establishments cause physical pollution. Power plants, steel, paper, car, plastic, and packing factories, which are big industrial establishments, throw environmentally hazardous solid and liquid substances. These substances are mostly toxic like arsenic, phenol, cyanide, chromium, and cadmium [24].

The chemical pollution of water began to cause critical health problems. It is estimated that in the future one of the most important water pollution problems will be the pollution caused by chemicals. Main metals causing chemical pollution are copper, zinc, mercury, nitrate, and phosphate [23].

Radioactive pollution in water can result from research agencies, hospitals, and some industrial fields. Radioactivity increases because of testing nuclear weapons. Therefore, rain water is getting dirty and, as a consequence, surface water is exposed to radioactive pollution [25].

Water might also be polluted by some pathogenic bacteria, parasites, and viruses that can originate from humans and animals [6]. Microorganisms that cause water pollution in terms of hygiene generally originate from diseases or human and animal excrements and urine being a porter [26]. An increase in bacteria population leads to bacterial pollution as a result of the decomposition of organic substances that are accumulated in the sea and inland waters or mixing of various pollutants apart from sewerage [25]. In stored waters, there is a lot of bacterial communities like the members of *Pseudomanas* sp., *Micrococcus* sp., *Achromobacter* sp., *Streptomyces* sp., and especially *Enterobacteriaceae*. The members of *Enterobacteriaceae* do not reproduce in clean water and their natural habitat is not potable water. Coliform bacteria are important in terms of human health and as an indicator of water pollution [24]. Their presence indicates that stool is mixed up with water through sewage directly or indirectly in one or more phases starting with the raw material to the transfer of water [27].

When it comes to water pollution, microbial pollution comes to mind first, even though there are many reasons of pollution. The most important of them is the pollution that arises from domestic and industrial wastes. Since domestic wastewater contains sewage waters and detergents, it also causes the indirect microbial and chemical pollution.

5. Wastewaters and classification

Wastewaters are formed as a result of the pollution of water used in households and industrial establishments [28].

For the waters that are disposed by being used in households or industry, "wastewater" definition is used. Wastewaters demonstrate biological, chemical, and physical pollution. While biological pollution consists of bacteria, fungi, parasites, and virus particles, and chemical pollution consists of toxic substances, decomposed organic substances, and phosphor, physical pollution consists of color, scent, foaming, temperature increase, and suspended matters. Heavy metals contain colorants that belong to the group of chemical pollutants and include industrial wastes and some pesticides [29].

5.1. Classification of wastewaters

Wastewaters are classified into two groups as domestic and industrial.

5.1.1. Domestic wastewaters

Wastewaters that originate from the dirty water from households and workplaces and do not include the industrial content of factories are called domestic wastewaters. Although their pollution rate is low they contain a high level of oily compounds, proteins, particles, chemical oxygen demand (COD), and detergents. For this reason, domestic wastewaters have a complex structure (**Table 1**) [30].

Domestic wastewaters are the waters that contain dirty looking and colorful soluble and insoluble matters from food wastes, kitchen lavabos, bathrooms, washing, and dishing machines and the matters that have organic and inorganic content and 99% of water [31].

Physical properties	Chemical components			Biological components
	Organics	Inorganics	Gases	
Solid matters	Carbohydrates	pH	Methane	Living cells
Heat	Oil and grease	Nitrogen	Oxygen	Plants
Color	Pesticides	Phosphorus	Hydrogen	Single cells
Smell	Phenols	Alkalinity	Sulfur	Viruses
	Proteins	Chlorides		
	Surface active agent	Heavy metals		
		Sulfur		
		Toxic components		

Table 1. Physical, chemical, and biological components of domestic wastewater [32].

http://dx.doi.org/10.5772/66188

Domestic wastewaters contain suspended, colloidal, and dissolved organic and inorganic substances. As well as this pollution arises from sewerages and detergents, it can also originate primarily from households and business enterprises. Moreover, domestic wastewaters contain pathogenic microorganisms such as bacteria, helminth, protozoa, and viruses. This situation increases the pollution rate of waters and indicates that water treatment is absolute. The indicator of the treatment's necessity is that some bacteria include R-plasmid. Since R-plasmid ensures antibiotic resistance to bacteria, untreated domestic wastewaters cause antibiotic-resistant bacteria to infect people and animals and create disease.

One of the most significant wastewaters that belong to domestic wastewaters is hospital-acquired wastewaters.

5.1.1.1. *Hospital-acquired wastewaters*

Hospital wastewaters contain micro and macro pollutants that come from various sources such as operating rooms, laboratories, investigation units, polyclinics, and drug use. The most important macro pollutants are bacteria and viruses while the most important micro pollutants are antibiotics, heavy metals (Hg, Pt, Gd, etc.), hormones and detergents/antiseptics. While microbiologic quality is determined for the usage of water, faecal pollution is identified by biologic and chemical indicators. In the content of biological indicators, there are total coliforms, faecal coliforms, faecal streptococcus, and *Clostridium perfringens*. Total coliforms are in the form of aerobic and facultative anaerobic, asporogenic and Gram-negative bacteria. Faecal coliforms, the marker of the pollution of water in which faecal coliforms and total coliform bacteria are found and which indicates the presence of pathogenic bacteria with human or animal excrements represent the presence of pathogenic bacteria and limited virus contamination. In hospital wastewaters, antibiotics such as ciprofloxacin, erythromycin, and sulfamethoxazole are found in high numbers in accredited adsorbable organic halide (AOX) and paracetamol. In municipal sewage, antibiotics such as ofloxacin and erythromycin are found in high numbers in AOX, paracetamol, and ibuprofen which is an analgesic [33].

In order for drugs to be stored longer and be easier to take, they must be quite durable and of high mobility quality in the liquid-phase while being produced. For this reason, active substances in drugs and biotransformation products lead to various factors by accumulating in the ecosystem. A lot of drugs such as antibacterial drugs, antibiotics, antifebrile, anodyne, synthetic steroids, cholesterol medicines, beta blockers and cytostatic drugs are the drugs detected in the ecosystem by studies performed [34].

Various drugs are used for various purposes during the treatment, protection, and development of human and animal diseases. These drugs cannot completely metabolize and they are removed from the body as they are or as a by-product in the form of ordure, urine, sweat, etc. [35].

In order for living beings to be treated, protected against microorganisms and infections and become resistant, many drugs should be taken. After the functions of these drugs in the body are over, they are removed through liver and kidneys. Medicine taken reaches the maximum level in the blood and when it starts to decrease the excretion also begins. While the excretion periods of drugs such as painkillers and antibiotics out of body are different, antibiotics are not removed for a long time. Drugs are removed from the body as urine, ordure, or metabolized product. In this way, they mix into wastewaters through the sewer system.

Drugs mix into wastewaters not only through excretion. With the disposal of unused drugs in households and hospitals, they also mix into wastewaters through the sewage.

The medicines found in wastewaters cannot be completely refined through the refinement. One of the biggest reasons for this situation is that hospital wastewaters directly mix with domestic waters without pretreatment. This affects primarily potable waters, underground waters, lakes, and rivers in a negative way.

Medicine remnants that mix into the potable water as a result of the inadequate refinement of domestic wastewaters negatively affect living beings in many ways. This effect arises especially from antibiotics. Antibiotics that enter the body through water cause pathogenic microorganisms to become resistant.

Due to the negative outcomes on living beings, hospital wastewaters should be refined before they are transferred to domestic wastewaters.

5.1.2. Industrial wastewaters

Pollution in the environment that originates from unavailable or economically unvalued wastes in the industrial system is called industrial pollution. The accumulation of permanent and toxic organic substances in industrial wastewaters creates serious problems. The facts that these wastewaters are not discharged into the receiving environment, pollutants are not biodegraded, and they have a toxic influence upon living beings create many troubles [36].

Industrial wastewaters comprise of various resources such as refrigerant waters, process wastewaters, and domestic qualified wastewaters. Because of this content, the refinement of industrial wastewaters becomes crucial [11].

Since industrial wastewaters contain heavy metal content, they create the most crucial environmental problem of the present day. Wastewaters containing heavy metals are the waters that are generally acidic and have a low biochemical oxygen demand (BOD) value. Aquatic life is affected by mixing of wastewaters into the receiving environment. Because of this situation, expensive refinement systems are needed in order to use water resources as potable water sources. Heavy metals contained in wastewaters make the mud impossible to use for agricultural purposes by affecting the refinement efficiency of domestic wastewaters. For this reason, the discharge of industrial wastewaters with heavy metal content into the sewer system has an important role [37].

5.2. Wastewater treatment

The treatment of domestic wastewaters takes place in three stages as mechanic, biologic, and chemical.

Physical treatment covers the refinement of solid matters in wastewaters. This treatment stage comprises of four units as grid/sieve/grinder, sand catcher/oil slinger, preliminary settling, and flotation [38].

The biologic treatment contains the stage in which organic matters contained in wastewaters are refined. This treatment happens along with the decrease in the organic matter amount by using and decomposing of organic substances as a nutrition substance by microorganisms. Domestic wastewater generally decreases nutrition and organic substances such as nitrogen and phosphorus contained in it. Biologic treatment helps microorganisms such as fungi, algae, protozoa, and metazoans and organisms belonging to bacteria and archaea. The most used processes in biologic treatment are activated sludge processes, air-conditioned lagoons, trickling filters, revolving biodiscs, and stabilization pools. Basic operations of this treatment are nitrification, denitrification, dephosphorization, waste stabilization and eliminating organics which are measured especially as BOD_5 and COD in wastewater [39, 40, 38].

Micropollutants that are common in water resources cannot be effectively eliminated with the present treatment systems and environmental impacts in the receiving environment. In particular, antibiotics and pharmaceutics are released into the environment after their production and consumption so they create a threat in the receiving environment. Conventional treatment processes, primarily biologic treatment systems, remain insufficient in the elimination of antibiotics. In order to remove antibiotics that are resistant to biodegradation, advanced oxidation processes with a high oxidation potential should be used [41].

In wastewater treatment establishments, antibiotics are generally removed from the environment by biodegradation and sorption with activated sludge. Antibiotic-resistant bacteria spread in nature thanks to the removing methods of activated sludge containing antibiotic-resistant organisms such as agricultural practices or burying into the pit [42].

Antibiotics are used to help the growth of animals along with the treatment of human and animal diseases. Antibiotics that enter the body are removed without being metabolized at the rates reaching 90%. For this reason, the main source of antibiotic pollution in nature is the antibiotics in human and animal faeces. In recent studies, it is determined that antibiotics are found in animal faeces and domestic wastewater sewage sludge besides various compartments. By regarding physical and chemical properties, antibiotics can reach sediments, soil, and underground water. It is determined that conventional treatment methods remain insufficient in the removal of low concentrated antibiotics in water. The high concentration of antibiotics in the environment causes the degradation of ecological balance by creating a toxic effect on microorganisms, and their low concentration causes pathogenic and nonpathogenic bacteria to gain antibiotic resistance. For this reason, in order to remove antibiotic pollution, alternative treatment methods are necessary [43].

Gao et al. [44] determined 14 antibiotics in total in the wastewater, and 18 antibiotics in the activated sludge in their study. In the activated sludge, floroquinons, and ofloxacin were determined at the highest rate. Wastewater treatment establishments cannot remove antibiotics completely and the removal rate ranges from 34 to 72%. The amount of antibiotics in water is determined to be higher in winter months in comparison with spring and fall months. At the same time, antibiotic remnants have an adverse effect on very different organisms in nature (they encourage reproduction). Because of the low treatment effect, wastewater treatment establishments are the major source of antibiotics in aquatic environments.

According to Li et al. [45], the removal activity of target antibiotics from water changed between 32 and 78% through conventional treatment. With the advanced treatment methods, the removal rate of target antibiotics became 85–100% and pollution probability of antibiotics decreased. In addition to this, in the risk assessment, the effects of ofloxacin and erythromycin on microorganisms in water are investigated by refining it. The majority of antibiotics cannot be absorbed or metabolized in the body. Moreover, the large part pass into the sewage system through urine and faeces and it comprises a significant part of the antibiotic source in nature.

In the study conducted by Zhang et al. [46], the elimination mechanism of three β-lactam, two fluoroquinolones, and two macrolide antibiotics was investigated in the wastewater treatment establishment, which has four different treatment methods among six wastewater treatment establishments in China, Dalian. In this study, fluoroquinolones and macrolide antibiotics were determined as dominant antibiotics at the exit of wastewater treatment establishment and in coastal waters. It is revealed that β-lactams are removed through biodegradation, for fluoroquinolones pretreatment is more effective than biologic treatment, and macrolide concentration increases dramatically after biological treatment. The reason for this is that macrolides that are covered by faeces particles are revealed [46].

Xu et al. [47] examined antibiotics and their resistant genes in a water treatment establishment in Beijing, China and the situation of the river into which water was discharged in their study. A total of 13 antibiotic resistance genes (ARGs) were examined. Sul-arg was found at the highest rate among all antibiotic resistance genes (ARGs). ARG quantity in the wastewater treatment establishment is higher than in the river. According to the correlation analysis, there is a positive relationship between tetracyclines and tetargs in water. This correlation could not be performed between sul-args and sulfonamides. A negative relationship was observed between the concentration of quinolone genes and enrofloxacin. When ARG abundance of the waters that are treated in the treatment establishment is examined, treatment establishment causes resistant genes to increase. Results show that treatment establishments have a function of a warehouse for resistance genes. As a result, treated water needs advanced treatment before it is sent to the natural aquatic environment. In the study, three antibiotic groups were studied as tetracycline, sulfonamides, and quinones that are known for their permanence in the aquatic environment. Tetracyclines are removed at the rate of 87.9% in sludge elimination establishments. In the elimination of tetracycline, biodegradation, and adsorption have an important role. In sludge elimination establishments, teta, tetm, tetw, and teto genes are the ones that are mostly found [47].

6. Antibiotics

Antibiotics are bioactive substances that kill or stunt the growth of the microorganism and have a high effect on synthetic or biological origin [48].

Antibiotics that are naturally obtained from plants and their extracts and are used for medical purposes have been brought into use as a result of Paul Ehrlich's studies in 1908. Paul Ehrlich revealed some chemical substances that are harmful to some bacteria and are less harmful to the host cells by investigating them [49].

http://dx.doi.org/10.5772/66188

Antibiotics are produced in nature by bacteria or fungi. The production of antibiotics by these living beings and their release into the environment result from their food competition with other species. Therefore, they produce antibiotics in their environment which extinguish other microorganisms or inhibit their growth. Antibiotics do not affect fungi, viruses, and protozoa since they are active only in bacterial infections. At the present time, antibiotics are produced synthetically. The microorganisms the production of which has provided the invention of antibiotics are fungi [50].

As antibiotics can be broad-spectrum affecting numerous bacteria, they can also be narrow-spectrum affecting limited bacteria. Furthermore, antibiotics with bactericide effect have an effect on bacteria by killing bacteria and antibiotics with bacteriostatic effect have an effect on bacteria by stopping their reproduction [51].

Although it has not been a long time since antibiotics have come into use, a rapid increase has been observed in their development. However, many problems have occurred during and after the consumption of these drugs. One of the main problems among them is bacterial resistance that develops against antibiotics.

There are a lot of reasons that bacteria develop resistance to antibiotics. The most important among them is that antibiotics are used without need and unconsciously.

Drugs that are used most frequently and in an excessive amount in the world are antibiotics. This usage also covers the unnecessary and unconscious use besides the proper use for treatment. The use for wrong purposes, misuse, and unnecessary use of antibiotics lead to bacterial resistance. For this reason, information about the usage of antibiotics should be given and the excessive and unnecessary usage should be prevented.

6.1. Classification of antibiotics

Antibiotics are separated into two groups according to their effect on microorganisms:

- **Classification according to antibiotic potencies**

1. Bacteriostatic: This type of antibiotics prevents the development and reproduction of bacteria without killing the cells.

2. Bactericide: This type of antibiotics destroys bacterial cells by causing heavy damage.

6.2. Mechanisms of action of antibiotics

6.2.1. The ones that inhibit cell wall synthesis

Bacteria are prokaryote microorganisms. They do not have real nucleus but they have cell walls. Cell walls protect bacteria from the external environment and antimicrobials. Cell wall contains pores 1–2 nm in diameter that is convenient to the transition of substances found in the external environment and nonselective. In short, they are not semipermeable. The transition of antimicrobials depends on the structure of the cell wall and molecular size of the drug.

Human cells have no cell wall. Thus, antibiotics (Penicillins and Beta-lactams) in this group cannot spoil the adhesion of human cells. These antibiotics affect either by adhering to Penicillin-Binding Proteins (PBP) or by spoiling the synthesis of cell wall without adhering to PBP (**Table 2**) [51].

The ones that inhibit the cell wall synthesis	**Beta-Lactams:**
	Penicillines
	Cephalosporins
	Monobactams (Aztreonam)
	Carbapenems (imipenem, Meropenem)
	Cycloserine
	Ristocetin
	Bacitracin
	Teicoplanin
	Vancomycin
The ones that inhibit cytoplasm membrane permeability	Polymyxins
	Gramicidin
	Nystatin
	Amphotericin B
	Candicein
	Ketoconazole and other antifungal imidazols
	Fluconazole and other antifungal trizols
	Hexachlorophene
	Cationic detergents
The ones that inhibit ribosome's protein synthesis	Tetracyclines
	Aminoglycosides
	Macrolides
	Amphenicols
	Lincosamides
	Fucidicasid
The ones that effect bacteria's genetic material break DNA and RNA	Fluoroquinolones
	Rifamycins
	Nalidixicasid
	Metronidazole
	Actinomycins

The ones that inhibit the cell wall synthesis	**Beta-Lactams:**
	Mitomycins
	Bleomycins
	Acyclovir
	Doxorubicin
	Daunorubicine
	Methotrexate
Bacterial antimetabolites	Sulfonamides
	Sulfones
	PAS
	Isoniazide (INH)
	Ethambutol
	Trimethoprim

Table 2. Classification of antibiotics [53].

6.2.2. The ones that inhibit the protein synthesis

These chemotherapeutic drugs are generally broad-spectrum and have a bacteriostatic effect. Tetracyclines which belong to this antibiotic group prevent the adhesion of t-RNA to ribosomes. As human ribosomes (60S + 40S) and bacterial ribosomes (50S + 30S) are structurally different, these antibiotics that show an effect by adhering to ribosomes do not affect human ribosomes and protein synthesis (**Table 2**) [51].

6.2.3. The ones that inhibit nucleic acid synthesis

The most important antibiotics which belong to this group are rifampicin and quinones. Rifampicin inhibits the transcription (RNA inhibition dependent on DNA). Quinones inhibit the formation of supercolid (DNA gyrase inhibitors).

Topoisomerases which are used in human DNA and RNA synthesis and enzymes which are used in the nucleic acid synthesis of microorganisms are different. For this reason, these antibiotics do not have a toxic effect on human cells (**Table 2**) [51].

6.2.4. The ones that increase cytoplasmic membrane permeability

These antimicrobials create an effect by splitting the membrane substances in bacteria, inhibiting sterol synthesis in fungi or spoiling the permeability by binding sterols.

The cytoplasmic membrane of human cell bears a resemblance with cytoplasmic membranes of bacteria and fungi. Therefore, these antibiotics can have a toxic effect on human cells when they are used in a systemic way (**Table 2**) [51].

6.2.5. *The ones with antimetabolic activity*

Antibiotics in this group are generally bacteriostatic. The ones that are broadly known are the drugs such as sulfonamides, sulfons, para-amino salicylic acid (PAS), ethambutols, and isoniazid. Sulfonamides and Sulfons stop the function of PAS and para-amino benzoic acid (**Table 2**) [51].

6.3. Basic antibiotic groups

6.3.1. *Beta-lactams*

Antibiotics containing beta-lactam circle which is found in the nucleus and is responsible for the antibacterial effect of molecules are called beta-lactam antibiotics. The beta-lactam circle is a saturated circle, which comprises one nitrogen and three carbons. Antibiotics in this group have bactericide effect by influencing the cell wall which consists of the murine of bacteria. Penicillin and Ampicillin are the most known antibiotics in the beta-lactam group [52].

6.3.2. *Vancomycin*

They have an effect on multiresistant bacteria [54]. These antibiotics inhibit cell wall synthesis by stopping RNA synthesis in bacteria, break the continuity of the peptidoglycan chain and spoil the cytoplasmic membrane structure. Vancomycin which has a narrow antibacterial spectrum affects Gram (+) cokes and *Clostridiums* [51].

6.3.3. *Tetracycline*

These antibiotics inhibit protein synthesis by adhering to 30S subunit of the microorganism ribosome. Tetracycline which affects both Gram (+) and Gram (-) bacteria is broad-spectrum and has bacteriostatic effect [55, 56]. Tetracycline affects numerous and various bacteria types. It is also effective against *Rickettsia* sp., *Chlamydia* sp., *Spirochaete* sp., *Mycoplasma* sp., *Leptospira* sp., and some protozoa [51].

6.3.4. *Aminoglycosides*

Aminoglycosides inhibit protein synthesis in ribosomes by adhering to 30S subunit of bacterial ribosomes. Moreover, they cause the misreading of genetic code that m-RNA has. These antibiotics are narrow-spectrum and have bactericide effect. They are effective only in aerobe bacteria as they are dependent on oxygen in the membrane cell [51].

6.3.5. *Macrolides*

These antibiotics inhibit protein synthesis that is dependent on RNA in bacteria. They provide this effect by preventing the continuity of the peptide chain and adhesion of t-RNA by binding 70S ribosome to 50S subunit. Bacteriostatic macrolides have an intense effect against Gram (+) cokes and bacillus [57].

6.3.6. *Chloramphenicol*

Chloramphenicol is the first broad-spectrum antibiotic. These antibiotics inhibit peptidyl transferase enzyme by binding bacterial ribosomes to 50S subunit and thus they inhibit pro-

http://dx.doi.org/10.5772/66188

tein synthesis in a reversible way. They are sensitive to coke, aerobe, anaerobe Gram (+) bacilli, and most of the Gram (-) bacteria. Furthermore, these antibiotics inhibit protein synthesis of bacteria in the tissue by transferring into the tissue [55, 58].

6.3.7. *Quinolones*

Quinolones affect bacteria by inhibiting DNA gyrase. This effect prevents DNA replication and creates bactericide impact. Moreover, the bacteria that are exposed to this antibiotic do not divide and die from stretching abnormally. They are effective in most Gram (-) bacteria and Gram (-) bacteria [59].

6.3.8. *Trimethoprim-sulfamethoxazole*

It is also known as cotrimoxazole. When Sulfamethoxazole (STX) is a sulfonamide, Trimethoprim (TMP) is a diaminopiriminid which inhibits bacterial dihydrofolate reductase competitively. They affect many Gram (+) and Gram (-) bacteria by causing unnoticeable synergistic bactericide effect when both drugs are used separately.

As a rule, the maximum synergistic activity of both antibacterial drugs, Trimethoprim (TMP), and Sulfamethoxazole (STX), occurs in bacteria types which are sensitive to both drugs. In the determination of the activity, sensitivity to TMP is more important [60].

6.4. Antibiotic resilience

Antibiotic resilience is simply the ability to resist against any antibiotic which spoils the reproduction function of a microorganism or causes its death. Resistance concerns the microorganism, patient, antibiotic, and environment or all of them. Resistance has no connection with virulence [61].

Antibiotic resistance spreads in three ways in bacteria:

1. Transfer of bacteria between people
2. Transfer of resistant genes between bacteria (generally through plasmids)
3. Transfer of resistant genes between genetic elements in bacteria [60]

The resistance that microorganisms show against antibiotics is classified in two groups as natural (phenotypic) and acquired (genotypic).

6.4.1. *Natural resistance*

Natural resistance is the situation that occurs when the microorganism cannot carry the structure affected by the drug as its quality or it cannot reach the target due to the structure of the drug. This resistance is not hereditary besides it is the key feature of bacteria and it is not related to the use of drugs.

For instance, microorganisms such as L-forms of bacteria and *Mycoplasma* that have no membrane have a natural resistance to antibiotics such as penicillin which inhibit the cell wall synthesis. Another example is that vancomycin cannot affect Gram (-) bacteria due to the fact that it cannot pass from adventitia [62].

6.4.2. *Acquired resistance*

Depending on the change in bacteria's genetic characteristic, it is the resistance which occurs as a result of taking DNA series that have resistance gene from another bacterium through transformation, transduction, or conjugation as it can be through mutations in a plasmid, chromosome, or transposon DNA. Furthermore, these bacteria can gain resistance against antibiotics to which they have been sensitive before [63]. Genetic originated resistance is examined in two groups as chromosomal and extrachromosomal.

6.4.2.1. *Chromosomal resistance*

It occurs as a result of mutations which happen spontaneously in the bacterial chromosome. Spontaneous mutations arise from some physical or chemical factors. Consequently, structural changes occur in the bacterial cell. In this situation, changes can happen in the drug's target in the cell or permeability of the cell to the drug can decrease [62].

6.4.2.2. *Extrachromosomal resistance*

Bacteria have extrachromosomal resistance plasmids that are called extrachromosomal elements, transposons that are active elements found on the chromosomes and bring chromosomes new antibiotic resilience, integrons, and antibiotics.

6.4.2.2.1.+ *Plasmids*

The structures that can be inside bacteria or outside the chromosomes in the DNA structure, bring some qualities to these bacteria and keep these qualities under control genetically are called plasmids.

Plasmids can have virulence factors besides resistance genes against antimicrobics and heavy metals. Plasmids which have resistance genes are called R-plasmids. R-plasmids transfer the resistant gene package by passing into other bacteria through transformation, transduction, and conjugation. Thus, they provide the spread of resistance [64].

6.4.2.2.2. *Transposons*

Transposons are the structures which can settle in different places in the bacterial chromosome or can be transferred from chromosome to plasmid, from plasmid to plasmid, from plasmid to DNA or bacteriophage. These structures are DNA series found over the replicon like a chromosome, plasmid or bacteriophage as they cannot replicate by themselves. They have an active role in the spread of the multiple drug resistant isolates of transposons by revealing them in a short time [62, 65].

6.4.2.2.3. *Integrons*

Integrons are active DNA elements which have the ability to capture genes, which codify antibiotic-resistant genes in enteric bacteria, with specific recombination. These genes that are captured by integrons are called gene tapes. Gene tapes are active genetic elements which comprise of only one gene and recombination zone which is free, little-alkali and called the 59-base element. As well as these gene tapes may not present in integrons at all, there can be 100 of them [66].

6.4.3. Cross-resistance

It is the situation when some microorganisms are resistant both to some drugs and at the same time to other drugs that have a similar mechanism. This resilience can be seen between structurally similar drugs like erythromycin and kanamycin as it can be seen between completely different drugs like erythromycin and lincomycin [62].

7. Antibiotic resistance in aquaculture and agriculture

Antibiotic concentrations below curative doses cause antibiotic resistance in many patient groups especially in critically ill patients [67].

The emergence of antibiotic-resistant bacteria is seen as an important health problem. For, thousands of patients die because of resistant bacteria. All efforts are concentrated on the decrease of existing antibiotic-resistant bacteria and antibiotic usage [2].

Rapidly developing antibiotic-resistant bacteria force public health services and health centers. American Centers for Disease Control and Prevention and Food and Agricultural Organization stated that antibiotic resilience has seriousness over the world. According to the predictions, 700,000 people die because of antibiotic resistance in a year. With the changes in temperature and rain regime, climate-sensitive bacteria and diseases will increase and spread to new regions, consequently, the situation will worsen [68].

Determination frequency and antibiotic concentrations are generally higher in January and May. The reason for this is that low-flow and low-temperature conditions cause antibiotics to be trapped by sediments. Antibiotic quantities vary per region. The highest quantities are found in estuaries and places where sewage is disposed. Antibiotic usage is more than 100,000–200·000 tonnes over the world and more than 25,000 tonnes in China. 80–90% of these antibiotics are released into nature through human urine and faeces. Pharmacologically active compounds in animal manure are used as a fertilizer in agriculture, and in conclusion, these compounds are accumulated in soil or mix into surface or underground waters [69].

Antibiotics are used as an environmental pollutant, in the treatment of diseases in a broad sense, in the protection and treatment of diseases in veterinary, and as growth promotive in aquaculture and agriculture [42].

Veterinary drugs are used for the protection and treatment of animal diseases and are one of the important components of environmental pollution as a result of intensive agricultural and aquaculture actions. Veterinary drugs are among the potentials of chemical pollutants and they have a biological effect in low concentrations like other drugs. While the annual usage of veterinary antibiotics in the United States reaches 11,000 tonnes, China follows it by 6000 tonnes. These quantities contain not only drugs with therapeutic purposes but also antibiotics which are used to promote production. In Europe, France leads these rates with 1064 tonnes, Holland follows it with 514 tonnes, and England with 403 tonnes. The most used antibiotics are tetracyclines, sulfonamides, β-lactams, and macrolides. The presence of veterinary antibiotics in nature causes the emergence of antibiotic-resistant bacteria and nontarget microorganisms are affected by drinking potable water that contains antibiotic remnants or by consumption of animal or herbal foods that contain antibiotics. Mixing of veterinary drugs into nature may cause the development of single, multiple, and cross-resistance in pathogens, commensals, and nonpathogens. Most of the veterinary drugs are feebly absorbed in the animal intestine. The remaining large quantity is removed with faeces. A small combination of these drugs removed undergoes a change, conjugates with polar molecules or remains the same. Consequently, these drugs can be detected in natural environments such as animal manure, soil, surface, and underground water resources. The major source of veterinary drugs in nature is biological remnants and the usage of dirty animal faeces in fertilization [70].

The usage of wastewaters for agricultural and other purposes by treating them provides many advantages such as the formation of alternative water resources, prevention of the pollution of surface and underground waters, and reduction of fertilizer usage. However, along with its advantages, it also has negative effects on public health and the environment. In order to minimize these effects, risks that origin from pathogens and chemicals that emerge from the wastewater usage should be evaluated well [71].

Waters that are polluted in many ways are treated by many methods with the progress of technology. The usage of wastewaters as irrigation waters by putting them through pretreatment or delivering into the land is one of these methods. Causing soil pollution by water pollution occurs in this method. Wastewaters from various resources pollute the soil and they have various effects on soil pollution [72].

Domestic wastewaters can be used in forests, pastures, lawns by being pretreated. The removal of wastewaters by using them in the irrigation of lands in this way creates serious health problems. Moreover, bacteria and pollutants in wastewaters are harmful to human health by being absorbed by the soil and reaching underground waters when the buffering effect decreases [72].

In the sector of aquaculture, antibiotics are intensively used to treat fish and protect it from diseases. Antibiotics that are applied to fish cause fish pathogens and zoonotic fish bacteria to gain resistance to antibiotics. Zoonotic fish bacteria which develop antibacterial resistance create danger for people and cause infections that are hard to treat [73].

The misuse of antibiotics affects human health directly or indirectly and complicates the treatment of fish diseases. Its direct effect is that fish bacteria and zoonotic fish bacteria gain resistance. These strains which are resistant create refractory infections when they infect people. The indirect antibiotic resilience occurs with the transfer of resistance plasmids in bacteria to human pathogens. In this way, human pathogens that gain resistance create resistant infections in people. Also in the studies conducted, it is revealed that multiple antibiotic resistance genes are transferred from fish pathogens to human pathogens [73].

In August 2011, 20 antibiotics that were taken from 20 different samples' regions taken from sediment and aquatic organisms in Dalian coastline were examined. Tetracyclines are dominant antibiotics in sea water. Sulfonamides are dominant antibiotics in sediment and aquatic organisms. Industrial aquaculture is the most significant reason for the pollution of coasts in developed and developing countries because of the intensive antibiotic usage. Antibiotic usage in China comprises the quarter of antibiotic usage over the world [74].

The state of 37 antibiotics was examined on 6 aquaculture farms around Hailing Island. Sulfamethoxazole, salinomycin, and trimethoprim were detected at the highest rate in water; ox tetracycline was detected at the highest rate in shrimp larva pools, enrofloxacin was detected at the highest rate in feed samples, and erythromycin was detected at the highest rate in sediment [75].

Wastewater usage in agriculture and land irrigations can be described as wastewater recycling. This usage brings many problems even if it is very economical. The usage of wastewater in agricultural activities by pretreating it is not enough to eliminate these results. Especially, domestic wastewaters constitute an enormous danger because of their content. Sewage and hospital wastewaters are the reasons for this danger. Antibiotics cause infections and antibiotic resistance in people besides the fact that they decrease the productivity in agriculture with the bacterial and parasite microorganisms they contain.

In aquaculture, which is a method used in fish farming, it is possible that bacterial and parasite infections occur. Therefore, antibiotics are used for the treatment and protection from infections. Antibiotic usage complicates treatment as well as it creates antibacterial resistance in fish and people.

8. Suggestion

Urbanization, an increase in industry and population, increases the water demand with each passing day. The most important need is water and nutrition's existence depends on water. For this reason, the amount of water, as well as its presence, is important for living beings.

The increase of water usage, its unconscious use, the involvement in pollutant activities, and not taking precautions against pollution have a negative influence on water amount. In this situation, people's awareness should be increased, and pollution pretending precautions should be taken. Besides these situations, water reutilization can be provided by treatment.

The reutilization of wastewater by treatment increases water amount and creates some sources for the use of living beings. This brings positive situations as well as many negative situations along with it.

Wastewater causes pathogenic factors in living beings because of its content as well as it causes antibiotic-resistant bacteria and the spread of pathogen microorganisms.

There are many ways for antibiotic resistance to occur and spread. The leading factors among them are the excessive and unconscious usage of antibiotics, the usage of broad-spectrum antibiotics, its accumulation in sewers through taking it to the body and urinating it, especially giving hospital sewer's accumulation to treatment facilities without its pretreatment.

The result of the insufficient treatment of wastewater treatment facilities is that the amount of antibiotics remains and an increase occurs as well. Due to the chemical structure of antibiotics, it may not come to light before entering the wastewater treatment. Antibiotics have the tendency to hold on to sediments due to their structure. In the stages of wastewater treatment, as a result of the decomposition of sediments, antibiotics come out. In this situation, the problems of not treating antibiotics arise. The usage of these waters for agricultural purposes also causes antibiotic resistance to spread.

Antibiotics are used for the purposes of treating diseases in humans and also for the same purposes in animals. In this situation, this can cause the emergence and spread of antibiotic-resistant bacteria.

To prevent the emergence and spread of antibiotic-resistant bacteria, first the awareness of people of antibiotic use should be raised. The usage of antibiotics in human and animal treatment should be reduced. Other waters that belong to the group of all sewage and wastewater, especially hospital sewage, should be pretreated before being discharged to wastewater treatments with biological treatment. Mechanic, chemical, and thermal treatment processes are included in pretreatment. Many pretreatment processes such as the process of oxidation, thermophilic pretreatment, sludge disintegration, ozonation, photocatalytic pretreatment, physicochemical pretreatment, and ultrasonic method should be used. Since wastewater treatment systems used fall short in some cases, new systems and pretreatment systems for antibiotic treatment should be developed.

Author details

Sadik Dincer* and Esra Sunduz Yigittekin*

*Address all correspondence to: sdincer@cu.edu.tr and esra-gokyuzu@hotmail.com

Biology Department, Science and Letter Faculty, Cukurova University, Adana, Turkey

References

[1] Polat A. Su Kaynaklarının Sürdürülebilirliği İçin Arıtılan Atıksuların Yeniden Kullanımı. Turkish Journal of Scientific Reviews. 2013; 6 (1): 58–62. (in Turkish)

[2] Centner TJ. Recent government regulations in the United States seek to ensure the effectiveness of antibiotics by limiting their agricultural use. Environment International. 2016; 94: 1–7.

[3] Akin M, Akin G. Importance of water, water potential in turkey, water basins and water pollution. Ankara University Faculty of Language, History and Geography Journal. 2007; 47, 2:105–118.

[4] Acikgoz S. Hydrogeochemistry of streams discharging to Sapanca Lake [thesis]. Kocaeli: Kocaeli University; 2008.

[5] Mutluay H, Demirak A. Water Chemistry. 1st ed. Istanbul: Beta; 1996. 140 p.

[6] Akturk S. Adana determination of microbial quality in water from Adana-Tufanbeyli road line [thesis]. Adana: Cukurova University; 2009.

[7] Oner M. Microbial Ecology. 1st ed. Izmir: Ege University Press; 1987. 282 p.

[8] Sunter AT. İcme ve Kullanma Sularinin Aritilmasi ve Dezenfeksiyonu. 6th National Sterilization Disinfection Congress; 1–5 April 2009; Samsun. 425–438 pp. (in Turkish)

[9] Curtis, H, Barnes NS. Biology. 5th ed. New York: Worth Publishers; 1989. 987 p.

[10] Murray RK, Granner DK, Mayes PA, Rodwell VW. Harpers's Biochemistry. 24th ed. Stamford, CT: Appleton & Lange Publishers; 1996. 868 p.

[11] Kamisli S. Determination of heavy metal and antibiotic resistance to gram-negative bacteria isolated from waste water treatment plants [thesis]. Adana: Cukurova University; 2014.

[12] Mengu GP, Akkuzu E. Global water crisis and water harvesting techniques. Journal of Adnan Menderes University Agricultural Faculty. 2008; 5(2):75–85.

[13] United Nations World Water Assessment Programme. The World Water Development Report 1: Water for People, Water for Life. Paris: Berghahn Books; 2003. 36 p.

[14] Muslu VA. Water pricing in the world and Turkey [specialist thesis]. Ankara: Ministry of Forest and Water Management; 2015.

[15] UNESCO. Water use. [Internet]. 2008. Available from www.unesco.org/water/iyfw2/water_use.shtml.

[16] Atalik A. Küresel isinmanin su kaynaklari ve tarim üzerine etkileri. Bilim ve Ütopya.2006; 139:18–21. (in Turkish)

[17] Dagli H. İçmesuyu kalitesi ve insan sağlığına etkileri. Bizim İller, İller Bankası Aylık Yayın Organı. 2005; 3: 16–21. (in Turkish)

[18] Haviland WA. Cultural Anthropology. 10th ed. Wadsworth Publishing Company; 2002. 576 p.

[19] Anonymous. [Internet]. 2014. www.cevreonline.com/su/dunyada%20suyun%20dagilimi.html.

[20] Gormez K. Cevre Sorunlari. 3rd ed. Ankara: Nobel Publication Distribution; 2015. 188 p.

[21] Ozkan A. Transboundary environmental damages in the Black Sea Basin and regional liability Regime. International Conference on Eurasian Economies; 17–18 September 2013; Russia. pp. 970–975.

[22] Arceivala SJ, Balman AH, Balman V. Wastewater Treatment for Pollution Control. Place of Publication not Identified. New Delhi; Ankara: Tata McGraw-Hill. 2002.

[23] Guler C, Cobanoglu Z. Su Kirliliği. 1st ed. Ankara: Çevre Sağlığı Temel Kaynak Dizisi; 1997. 47 p. (in Turkish)

[24] Ozarslan A. Determinetion of fecal coliform levels and antibiotics resistance frequency of Adana drinking water [thesis]. Adana: Cukurova University; 2009.

[25] Tanyolaç J. Limnology. Ankara: Hatipoğlu Publications; 1993. 235 p.

[26] Demirekin H. Sensitivity analysis of environmental problems in Isparta city [thesis]. Isparta: Suleyman Demirel University; 2001.

[27] Dincer S, Matyar F, Sonmez N. Seyhan nehrinin fekal kirlilik duzeyi ve fekal koliformlarin antibiyotik hassasiyetleri. 12. Biyoteknoloji Kongresi. 2001; Ayvalık, 252–255. (in Turkish)

[28] Samsunlu A. Wastewater Treatment. 3rd ed. Istanbul: Birsen Publisher; 2011. 647 p.

[29] San NO. Treatment of wastewater with heavy metal and reactive dye by *Rhodotorula sp.* [thesis]. Ankara: Ankara University; 2007.

[30] Gunes E. Anaerobic treatment of domestic wastewaters in upflow anaerobic sludge bed (UASB) reactor at temperate conditions and chemical post-treatment applications [thesis]. Istanbul: Istanbul Technical University; 2008.

[31] Dogan M, Saylak M. Su Kimyası. Kayseri: Erciyes University Publications; 2000. 120 p. (in Turkish)

[32] Metcalf E. Wastewater Engineering, Treatment and Reuse. 4th ed. New York: McGraw-Hill Education; 2003. 1771 p.

[33] Yasar A, Dogan EC, Arslan A. Macro and micro pollutants and treatment options in hospital wastewaters. Erciyes University, Journal of Institute of Science and Technology. 2016; *29*(2): 144–158.

[34] Saygi S. Cevre ve insan sagligi yönünden ilaç atiklarinin onemi. Marmara Pharmaceutical Journal. 2012; 16: 82–90. DOI: 10.12991/201216406. (in Turkish)

[35] Sonmez G, Isik M. Sulardaki Ilac Kalintilarinin Ileri Oksidasyon Yontemleri Ile Giderimi. Turkish Journal of Scientific Reviews. 2013; *6*(1):68–73. SSN: 1308-0040, E-ISSN: 2146-0132. (in Turkish)

[36] Turker C. Treatability of segregated textile effluents containing auxiliary chemicals [thesis]. Istanbul: Istanbul Technical University Institute of Science; 2013.

[37] Turkman A, Aslan S, Ege I. Lead removal from wastewaters by natural zeolites. Dokuz Eylul University Faculty of Engıneerıng Journal of Engıneerıng Scıence. 2001; 3:2:13–19.

[38] Sinan RK. Estimation of primary treatment and biological treatment effluent parameters by artificial neural networks in domestic wastewater treatment plants [thesis]. Selcuk University; 2010.

[39] Chojnacka K. Using bisorption to enrich the biomoass of Choleralla vulgaris with microelements to be used as mineral feed supplement. World Journal of Microbiology and Biotechnology.2007; 23:1139–1147.

[40] Ozturk I, Timur H, Koskan U. Principals of Wastewater Treatment: Municipal. Industrial Wastewater Treatment and Sludge Control. Ankara, Turkey: Ministry of Environment and Forestry; 2005.

[41] Ozkal CB, Pagano SB. Evaluation of antibiotics and antibiotic resistant bacteria removal by photo-catalysis. Nigde University Journal of Engineering Sciences. 2016; 5(1): 1–18.

[42] Li W, Shi Y, Gao L, Liu J, Cai Y. Occurrence, distribution and potential affecting factors of antibiotics in sewage sludge of wastewater treatment plants in China. Science of the Total Environment. 2013; 445: 306–313.

[43] Yalap KS, Balcioglu IA. Effects of water components on the advanced oxidation of a veterinary antibiotic, oxytetracycline. ITU Journal of Water Pollution Control. 2008; 18: 2–3.

[44] Gao L, Shi Y, Li W, Niu H, Liu J, Cai Y. (2012). Occurrence of antibiotics in eight sewage treatment plants in Beijing, China. Chemosphere. 2012; 86(6):665–671.

[45] Li W, Shi Y, Gao L, Liu J, Cai Y. Occurrence and removal of antibiotics in a municipal wastewater reclamation plant in Beijing, China. Chemosphere. 2013; 92(4): 435–444.

[46] Zhang H, Liu P, Feng Y, Yang F. Fate of antibiotics during wastewater treatment and antibiotic distribution in the effluent-receiving waters of the Yellow Sea, northern China. Marine Pollution Bulletin. 2013; 73(1): 282–290.

[47] Xu J, Xu Y, Wang H, Guo C, Qiu H, He Y, Meng W. Occurrence of antibiotics and antibiotic resistance genes in a sewage treatment plant and its effluent-receiving river. Chemosphere. 2015; 119: 1379–1385.

[48] Topal M, Senel GU, Topal EIA, Obek E. Antibiotics and usage areas. Journal of Institute of Science and Technology. 2015; 31(3):121–127.

[49] Basturk S. Escherichia coli, Klebsiella pneumoniae, Pseudomonas aerugino save Acinetobacter baumannii Suslarında Çesitli Kinolon Grubu Antibiyotiklerin Duyarlılıklarının Arastırılması [thesis]. T.C. Sağlık Bakanlığı Haseki Eğitim Ve Arastırma Hastanesi Enfeksiyon Hastalıkları Ve Klinik Mikrobiyoloji Kliniği İstanbul; 2005. (in Turkish)

[50] Hoel D, Williams DN. Antibiotics: past, present and future. Unearthing nature's Magic Bullets. 1997; 101(1):114–118.

[51] Gurses F. 2015. Treatability of antibiotic formulation effluents by fenton-like and photo-fenton-like advanced oxidation processes [thesis]. Istanbul: Istanbul Technical University; 2015.

[52] Oncul O. Antibiyotikler 1. İ.Ü. Cerrahpasa Tıp Fakültesi Sürekli Tıp Eğitimi Etkinlikleri, Akılcı Antibiyotik Kullanımı ve Eriskinde Toplumdan Edinilmis Enfeksiyonlar Sempozyumu. 2002; 31: 23–28. (in Turkish)

[53] Akkan AG. Antibiyotiklerin sınıflandırılması. I.U. Cerrahpasa Faculty of Medicine Continuing Medical Education Activities Antibiotics Practice Symposium; 2–3 May 1997; Istanbul. pp. 52–62.(in Turkish)

[54] Strohl WA, Rouse H, Bruce DF. Lippincott's Illustrated Reviews. Microbiology. Nobel Medicine Bookstores. 2006; 516:44–47.

[55] Alcamo EI. Fundamentals of Microbiology. 6th ed. Sudbury Massachusetts: Jones and Bartlett Publishers; 2001. 832 p.

[56] Lefever Kee J, Hayes ER, McCuistion LE. Pharmacology: A Nursing Process Approach. 7th ed. Printed in The United States America: Elsevier Saundres; 2012. 983 p.

[57] Allen N. Effects of macrolide antibiotics on ribosome function. In Macrolide Antibiotics. Birkhäuser Verlag, Basel. 2002; 261–280 pp.

[58] Braibant M, Gilot P, Content J. The ATP binding cassette (ABC) transport systems of Mycobacterium tuberculosis. FEMS Microbiology. 2000; 24: 449–467

[59] Algun U, Arisoy A, Gunduz T, Ozbakkaloglu B. The resistance of Pseudomonas aeruginosa strains to fluoroquınolone group of antibiotics. Indian Journal of Medical Microbiology. 2004; 22(2): 112–114.

[60] Rang HP, Dale MM, Ritter JM, Moore PK. 2006. Pharmacology. 5th ed. Edinburgh: Elsevier Churchill Livingstone; 2006. 797 p.

[61] Durupinar B. Antibiyotiklere dirençte yeni eğilimler. Klimik Journal. 2001; 14(2): 47–56. (in Turkish)

[62] Yuce A. Antimikrobik İlaçlara Direnç Kazanma Mekanizmaları. Klimik Journal. 2001; 14(2): 41–46. (in Turkish)

[63] Tanir G, Gol N. Antibiyotik Direnci. Klimik Journal. 1999; 2:12. (in Turkish)

[64] Gur D, Tutar I, Vardar Unlu G. (2001). İsepamisinin hastane izolatı Gram-negatif bakterilere karsı in vitro etkisi. Turkish Journal of Hospital Infections. 2001; 5(1):19. (in Turkish)

[65] Aygun G, et al. The antibiotic susceptibility patterns of *acinetobacter baumannii* strains isolated from nosocomial infections in intensive care unit. Journal of ANKEM. 2002; 16(1): 85–88.

[66] Roy PH. Integrons: Novel mobile genetic elements mediating antibiotic resistance in *Enterobacteria* and *Pseudomonas.* APUA Newsletter. 1995; 13(3): 1–4.

[67] Carlier M, Stove V, Wallis SC, De Waele JJ, Verstraete AG, Lipman J, Roberts JA. Assays for therapeutic drug monitoring of β-lactam antibiotics: A structured review. International Journal of Antimicrobial Agents. 2015; 46(4): 367–375.

[68] Centner TJ. Efforts to slacken antibiotic resistance: Labeling meat products from animals raised without antibiotics in the United States. Science of the Total Environment. 2016; 563: 1088–1094.

[69] Shi H, Yang Y, Liu M, Yan C, Yue H, Zhou J. Occurrence and distribution of antibiotics in the surface sediments of the Yangtze Estuary and nearby coastal areas. Marine Pollution Bulletin. 2014; 83(1): 317–323.

[70] Bártíková H, Podlipná R, Skálová L. Veterinary drugs in the environment and their toxicity to plants. Chemosphere. 2016; 144: 2290–2301.

[71] Kukul YS, Calıskan Unal AD, Anac S. Wastewater reuse in agriculture and health risks. The Journal of Ege University Faculty of Agriculture. 2007; 44(3):101–116.

[72] Tolunay D. Toprak kirlenmesi ve yanlıs arazi kullaniminin yarattigi sorunlar ile cozum onerileri. Journal of the Faculty of Forestry Istanbul University (JFFIU). 1992; 42(1–2):156–168. (in Turkish)

[73] Avsever ML, Turk N, Tunaligil S. The increase of antibiotic resistance in aquaculture and its effects on human health. research articles. The Journal of Bornova Veterinary Control and Research.2010; 32(46): 19–23.

[74] Na G, Fang X, Cai Y, Ge L, Zong H, Yuan X, Zhang Z. Occun, China. Marine Pollution Bulletin. 69(1): 233–237.

[75] Chen H, Liu S, Xu XR, Liu SS, Zhou GJ, Sun KF, Ying GG. Antibiotics in typical marine aquaculture farms surrounding Hailing Island, South China: occurrence, bioaccumulation and human dietary exposure. Marine Pollution Bulletin. 2015; 90(1): 181–187.

Chapter 6

Comparative Assessment of Pharmaceutical Removal from Wastewater by the Microalgae *Chlorella sorokiniana*, *Chlorella vulgaris* and *Scenedesmus obliquus*

Carla Escapa Santos, Ricardo Nuno de Coimbra,
Sergio Paniagua Bermejo,
Ana Isabel García Pérez and Marta Otero Cabero

Additional information is available at the end of the chapter

http://dx.doi.org/10.5772/66772

Abstract

In view of risks associated with the discharge of pharmaceuticals in the aquatic environment, the objective of this work was to assess the removal of paracetamol, salicylic acid and diclofenac from water by a microalgae-based treatment. For a comparison purpose, the growth and kinetic parameters for the removal of drugs were determined for three different microalgae strains, namely *Chlorella sorokiniana, Chlorella vulgaris* and *Scenedesmus obliquus*. It was found that the drugs removal efficiency by these strains was related to their growth. Comparing the three pharmaceuticals, the salicylic acid was the most efficiently removed, especially by *S. obliquus* (>93% batch culture, >99% semicontinuous culture) and *C. sorokiniana* (>73% batch culture, >93% semicontinuous culture). Contrarily, paracetamol was the most poorly removed, the maximum efficiencies being those attained by *C. sorokiniana* (>67% batch culture, >41% semicontinuous culture). On the other hand, diclofenac was efficiently removed only by *S. obliquus* (>98% batch culture, >79% semicontinuous culture). For the three considered drugs, *C. vulgaris* was the strain showing the lowest removal capacity. The large differences here revealed between microalgae strains regarding their removal capacity of pharmaceuticals, pointed to the strain selection as a key issue for a successful application in wastewater treatment.

Keywords: emerging contaminants, wastewater treatment, phytoremediation, paracetamol, salicylic acid, diclofenac

1. Introduction

Emerging contaminants (ECs) include a wide range of compounds and may be defined as naturally occurring, manufactured or man-made chemicals or materials that have been found or are suspected to be present in various environmental compartments and whose toxicity or persistence are likely to significantly alter the metabolism of a living being [1]. Among them, pharmaceuticals have received considerable attention with respect to their environmental fate and toxicological properties over the last decade [2]. Pharmaceuticals represent an especially worrying class since they were designed to cause a physiological response and their presence in the environment may affect non-target individuals and species [3]. This concern on pharmaceuticals presence in the aquatic environment has led to the recent consideration by European regulations within the Water Framework Directive (2000/60/EC) (WFD). The Commission proposal of 31 January 2012 foresaw the inclusion of three pharmaceuticals, namely diclofenac, 17-beta-estradiol (E2) and 17-alpha-ethinylestradiol (EE2) in the list of priority substances. Instead, by the EU Decision 2015/495, these compounds together with another estrogen (E1) and three antibiotics (azithromycin, clarithromycin and erythromycin) were finally included in the first watch list of substances to be monitored in all member states to support future reviews of the priority substances list [4].

Pharmaceuticals in domestic sewage or from hospital or industrial discharges end in municipal sewage treatment plants (STPs), but conventional wastewater treatments have been reported to be ineffective in the removal of such pollutants, with efficiency values of <5 to 40% [5]. In fact, STPs were not originally designed for the removal of pharmaceuticals due to the non-existence of limiting regulations on their discharge [6, 7]. Consequently, STPs are important sources of such pollutants in the aquatic environment [8, 9]. In this regard, Verlicchi et al. [10], who reviewed the occurrence of 118 pharmaceuticals in the influent and effluent of 244 STPs, found that the occurrence of some of them in the effluent discharged into surface water bodies may pose a medium-high (acute) risk to aquatic life. Among the studied pharmaceuticals, diclofenac was shown to have the highest average mass load (240 mg/1000 inhabitant) in the effluents of municipal STPs [10]. The removal efficiencies of diclofenac in conventional STPs have been reported to be about 17% [11], which translates into relative high concentrations in the corresponding effluents [12].

In the recent years, phytoremediation of waters by using photoautotrophic aquatic organisms such as algae has gained attention for the removal of both organic and inorganic pollutants [13–15]. Microalgae are characterized by high photosynthetic efficiency, high growth rates, wide adaptability and high potential to remove inorganic nutrients from the wastewater. The principal mechanism of algal nutrient removal is their uptake into the cell biomass [16]. The main advantages of using microalgae for nutrients removal during the tertiary treatment of wastewaters are the possibility of recycling the assimilated nitrogen and phosphorus into algal biomass as a fertilizer, as a source of products (e.g. paraffin, olefin, glycerol, protein, anti-oxidant, pigment, plastic, etc.), or as biofuel, and also the generation of an oxygenated high-quality effluent [17]. However, although the capability of microalgae wastewater treatments systems to remove organic matter and nutrients has been deeply studied, little is known about the removal of ECs, such as pharmaceuticals, by algae. In fact, it has already been claimed the necessity of further studies on the removal of this sort of pollutants by algal systems [18].

In this context, the aim of this study was to determine and compare the potential of green microalgae *Chlorella sorokiniana, Chlorella vulgaris* and *Scenedesmus obliquus* to remove paracetamol, salicylic acid and diclofenac from water. The strains used in this work were selected since they are known to have fast growth rates and potential for wastewater treatment due to their tolerance to the severe environmental conditions found in municipal wastewater and some industrial wastewaters [19].

2. Materials and methods

2.1. Microorganisms and culture conditions

The microalgae strains used in this study were *C. sorokiniana* CCAP 211/8 K from UTEX Culture Collection of Algae, *C. vulgaris* SAG 221-12 from SAG Culture Collection of Algae and *S. obliquus* SAG 276-1 from SAG Culture Collection of Algae. These microalgae strains are among the most commonly used for wastewater treatment have high growth rates and are able to grow under a wide range of conditions [19], which motivated their choice for this study.

The inoculum of each strain was cultivated in 250-ml Erlenmeyer flasks in the standard culture medium Mann and Myers [20], which is composed of (per litre of distilled water): 1.2 g $MgSO_4.7H_2O$, 1.0 g $NaNO_3$, 0.3 $CaCl_2$, 0.1 g K_2HPO_4, 3.0 x 10^{-2} g Na_2EDTA, 6.0 x 10^{-3} g H_3BO_3, 2.0 x 10^{-3} g $FeSO_4.7H_2O$, 1.4 x 10^{-3} g $MnCl_2$, 3.3 x 10^{-4} g $ZnSO_4.7H_2O$, 7.0 x 10^{-6} g $Co(NO_3)_2.6H_2O$, 2.0 x 10^{-6} g $CuSO_4.5H_2O$. The inoculum was kept inside a vegetal culture chamber, where growth occurred under controlled temperature (25 ± 1°C), irradiance in the range of photosynthetically active radiation (175 $\mu E\ m^{-2}\ s^{-1}$), photoperiod (12:12) and shaking (250 rpm).

Bubbling column photobioreactors (PBRs) with spherical bases (40 mm diameter and 300 mm height with 300 ml capacity) were used for the experimental setup, keeping an operating volume of 250 ml. In each PBR, the Mann and Myers culture medium was inoculated with the required volume of the corresponding pre-cultured microalgae in order to have an initial concentration of about 3×10^6 cells ml^{-1}.

During the experimental phase, the culture was aerated with filtered air (0.22-μm sterile air-venting filter, MillexFG50-Millipore), at a rate of 0.3 v/v/min, enriched with CO_2 at 7% v/v, which was injected on demand to keep a constant pH (pH = 7.5 ± 0.5), as controlled by a pH sensor. The irradiance supplied during this phase was 370 $\mu E\ m^{-2}\ s^{-1}$, which was provided by eight fluorescent lamps (58 W, 2150 lumen, Philips, France). The photoperiod was maintained in 12:12 h light/dark and the temperature in 25 ± 1°C.

2.2. Experimental setup

PBRs were operated in batch mode until the end of the exponential growth phase and then under semicontinuous mode till the growth parameters remained constant at the steady state.

During the batch culture, an aliquot of 5 ml was daily taken from each PBR for the analytical determinations, this volume being replaced with distilled water to keep the operation volume. During the semicontinuous culture, 30% of the culture volume was daily harvested and used for analysis, this volume being replaced with fresh medium.

For each strain of microalgae used in this work (*C. sorokiniana, C. vulgaris* and *S. obliquus*), three treatments were conducted: (i) a treatment with inoculated culture medium and 25 mg l^{-1} paracetamol (with *C. sorokiniana* PCS, *C. vulgaris* PCV, *S. obliquus* PSO), (ii) a treatment with inoculated culture medium and 25 mg l^{-1} salicylic acid (with *C. sorokiniana* SCS, *C. vulgaris* SCV, *S. obliquus* SSO) and (iii) a treatment with inoculated culture medium and 25 mg l^{-1} diclofenac (with *C. sorokiniana* DCS, *C. vulgaris* DCV, *S. obliquus* DSO). Also, the corresponding positive controls with inoculated culture medium (with *C. sorokiniana* CCS+, *C. vulgaris* CCV+ and *S. obliquus* CSO+) were run. The negative controls consisted of 25 mg l^{-1} paracetamol (CP−), salicylic acid (CS−) or diclofenac (CD−) in culture medium with no microalgae. For each strain, experiments were run in triplicate and under identical conditions in all the PBRs. Paracetamol ($C_8H_9NO_2$, ≥99%) was supplied by Roic Pharma, salicylic acid ($C_7H_6O_3$, ≥99%) by Panreac and diclofenac ($C_{14}H_{10}C_{12}NNaO_2$, ≥99%) by Sigma-Aldrich.

Throughout the experiments, the growth of the culture was daily monitored by the determination of biomass concentration and cell density. The removal of pharmaceuticals was daily determined by the analysis of the remaining concentration of this drug in the culture medium. All analyses were conducted in triplicate.

2.3. Analytical methods

Biomass concentration (Cb) was determined by optical density at 680 nm (OD_{680}) by spectrophotometric (UV/visible spectrophotometer BECKMAN DU640) and verified by dry weight. Preliminary studies were conducted to determinate the relationship between dry weight and OD_{680} for each strain; as shown in Eq. (1) for *C. sorokiniana*, in Eq. (2) for *C. vulgaris* and in Eq. (3) for *S. obliquus*:

$$OD_{C.S\,680} = 5.1834 \times C_b + 0.0128, \quad R^2 = 0.9983 \tag{1}$$

$$OD_{C.V\,680} = 2.7933 \times C_b + 0.0317, \quad R^2 = 0.9958 \tag{2}$$

$$OD_{S.O\,680} = 2.0098 \times C_b + 0.0451, \quad R^2 = 0.9915 \tag{3}$$

Dry weight measurements were performed by filtering 10 ml of culture through a 0.45 μm Whatman filter, which was then washed with 20 ml HCl (0.5 M) to dissolve precipitated salts. Then, the filtrate was dried in an oven at 80°C for 24 h. Additionally, the growth of the culture was measured as cell density (Nc) by cell counting with a Neubauer chamber.

The initial and remaining pharmaceuticals concentration in the culture medium was quantified by a Waters HPLC 600 equipped with a 2487 Dual λ Absorbance Detector. A Phenomenex Gemini-NX C18 column (5 μm, 250 mm × 4.6 mm) was used for the separation. The wavelengths of detection were 246 nm for paracetamol, 236 nm for salicylic acid and 276 nm for diclofenac. The mobile phase consisted of a mixture of acetonitrile to water (30:70, v/v) for the analysis of paracetamol and a mixture of acetonitrile to water to orthophosphoric acid (70:30:0.1, v/v/v) for salicylic acid and diclofenac. HPLC quality acetonitrile (CH_3CN) and orthophosphoric acid (H_3PO_4) from Prolabo Chemicals and ultrapure water obtained by a Millipore System were used for the preparation of the mobile phase. Before use, each mixture was passed through a Millipore 0.45-μm pore-size filter and degasified in an ultrasound bath for 30 min. Before analysis, all the samples were centrifuged twice at 7500 rpm for 10 min (SIGMA 2-16P centrifuge). For the chromatographic analysis, the mobile phase flow rate was 1 ml min^{-1} and the injection volume was 100 μl.

2.4. Data analysis

Growth kinetics were resolved in OriginPro 8 using the classic model originally described by Verhulst [21] called logistic model, which has been proved to fit the growth of microalgae [22]. The logistic model fits to a sigmoidal curve that describes the relationship between microorganisms' growth and density in limited environmental conditions (Eq. (4)).

$$N = \frac{K}{1 + e^{a-rt}} \qquad (4)$$

Where N (g l^{-1}) is the algal density at time t (h), K (g l^{-1}) is the carrying capacity (the maximum algal density reached in the culture), a is a constant in the logistic model that refers to the relative position from the origin and indicates the duration of the lag phase and r (d^{-1}) is the specific growth rate.

Furthermore, the kinetic curves for the removal of pharmaceuticals were fitted to the logistic model. In each case, the parameter K (g l^{-1}) is the maximum removal capacity by the microalgae in the culture. The parameter a is a constant in the logistic model that indicates the delay in the beginning of the target compounds removal and the parameter r (d^{-1}) is the specific removal rate.

Finally, differences among the strains with respect to the kinetic parameters of growth and removal of pharmaceuticals were compared by a non-parametric test using IBM SPPS Statistics 21. The comparison of means was performed by means of the U Mann-Whitney test. Significance was defined at $p \leq 0.05$.

For the removal of pharmaceuticals, the volumetric efficiency for each target compound was calculated as the difference between its average concentration in the influent (C_{inf}) and in the effluent (C_{efflu}) at every sampling day, considering the daily dilution rate of the corresponding operation stage (D) (Eq. (5)). During the batch culture these efficiencies were cumulatively expressed as milligram per litre and as milligram per litre per day during the steady state of the semicontinuous culture:

$$\text{Volumetric efficiency} = \left(C_{\text{inf}} - C_{\text{efflu}}\right) \times D \tag{5}$$

The specific efficiency of the removed pharmaceuticals was calculated as the ratio between the volumetric efficiency and the biomass concentration (Cb) (Eq. (6)). Likewise, during the batch culture these efficiencies were cumulatively expressed as milligram per gram per biomass and as milligram per gram day during the steady state of the semicontinuous culture:

$$\text{Specific efficiency} = \frac{\left(C_{\text{inf}} - C_{\text{efflu}}\right) \times D}{C_b} \tag{6}$$

3. Results

3.1. Growth of the culture

The growth curves of *C. sorokiniana*, *C. vulgaris* and *S. obliquus* during the batch culture, of either the treatments or the controls, showed a typical sigmoidal growth of 8–10 days until reaching the steady state. On the other hand, during the semicontinuous mode, daily dilution rates produced instability and the growth rate declined throughout several days until the growth parameters remained constant during the steady state. This instability is a typical behaviour in the microalgae culture when the growth conditions change and it is related with an adaptation phase (**Figures 1–3**).

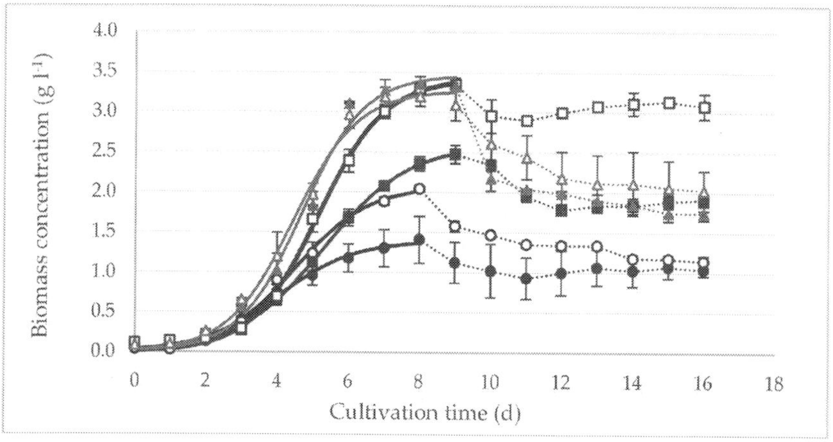

Figure 1. Growth curves of *C. sorokiniana* (CCS+ ●, PCS ○), *C. vulgaris* (CCV+ ■, PCV □) and *S. obliquus* (CSO+ ▲, PSO △) for the paracetamol treatments. Dots correspond to experimental data and continuous lines correspond to fittings by the logistic kinetic model during batch culture. Experiments were performed in triplicate and bars show standard derivations. Note: experimental points obtained during semicontinuous culture are connected with dashed lines.

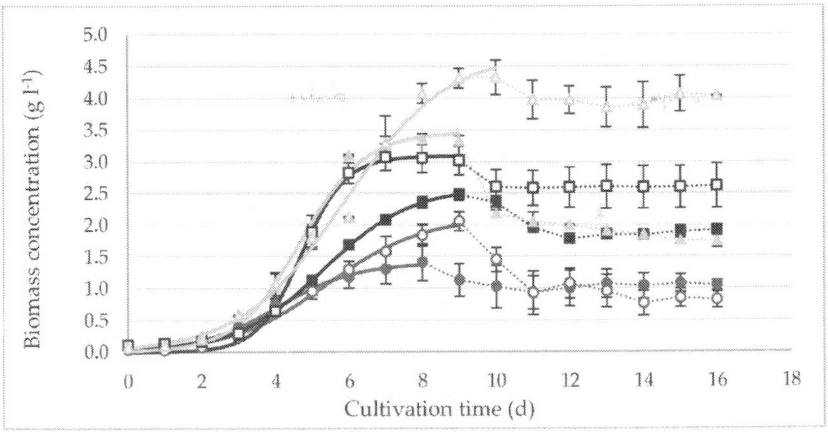

Figure 2. Growth curves of *C. sorokiniana* (CCS+, ●; SCS, ○), *C. vulgaris* (CCV+, ■; SCV, □) and *S. obliquus* (CSO+, ▲; SSO, △) for the salicylic acid treatments. Dots correspond to experimental data and continuous lines correspond to fittings by the logistic kinetic model during batch culture. Experiments were performed in triplicate and bars show standard derivations. Note: experimental points obtained during semicontinuous culture are connected with dashed lines.

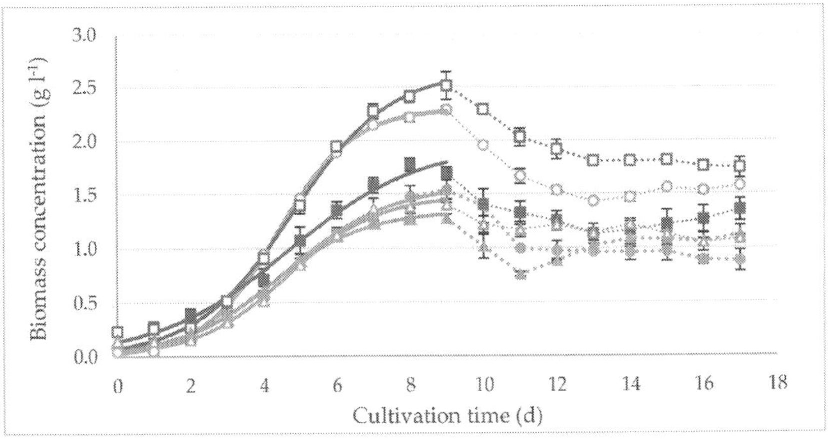

Figure 3. Growth curves of *C. sorokiniana* (CCS+, ●; DCS, ○), *C. vulgaris* (CCV+, ■; DCV, □) and *S. obliquus* (CSO+, ▲; DSO, △) for the diclofenac treatments. Dots correspond to experimental data and continuous lines correspond to fittings by the logistic kinetic model during batch culture. Experiments were performed in triplicate and bars show standard derivations. Note: experimental points obtained during semicontinuous culture are connected with dashed lines.

3.1.1. *Growth of the culture under paracetamol addition*

The microalgae growth curves of *C. sorokiniana*, *C. vulgaris* and *S. obliquus* under the presence of paracetamol, the corresponding positive controls and their respective fittings to the logistic kinetic model are represented as values of biomass concentration versus time in **Figure 1**. The differences among the treatments were analysed according to growth kinetic parameters, as shown in **Table 1**.

	CCS+	PCS	CCV+	PCV	CSO+	PSO
Cb_0 (g l^{-1})	0.04	0.04	0.11	0.11	0.08	0.08
Nc_0 (cell ml^{-1})	3.20×10^6	3.20×10^6	1.21×10^6	1.21×10^6	8.35×10^5	8.35×10^5
Cb_m (g l^{-1})	1.41 ± 0.29	2.05 ± 0.03	2.48 ± 0.11	3.35 ± 0.08	3.33 ± 0.06	3.09 ± 0.20
Nc_m (cell ml^{-1})	$2.12 \times 10^8 \pm 0.49 \times 10^8$	$4.20 \times 10^8 \pm 0.22 \times 10^8$	$1.18 \times 10^8 \pm 0.09 \times 10^8$	$2.17 \times 10^8 \pm 0.20 \times 10^8$	$4.77 \times 10^7 \pm 0.01 \times 10^7$	$4.62 \times 10^7 \pm 0.20 \times 10^7$
a	3.77 ± 0.01	4.33 ± 0.21	4.47 ± 0.25	5.58 ± 0.03	5.45 ± 0.43	4.97 ± 0.31
K (g l^{-1})	1.40 ± 0.29	2.09 ± 0.02	2.60 ± 0.15	3.42 ± 0.15	3.46 ± 0.08	3.27 ± 0.15
r (d^{-1})	0.94 ± 0.06	0.96 ± 0.07	0.84 ± 0.06	1.08 ± 0.04	1.16 ± 0.07	1.12 ± 0.03
R^2	0.9935	0.9939	0.9971	0.9968	0.9874	0.9886

Cb_o, initial biomass; Nc_o, initial number of cells; Cb_m, maximum biomass; Nc_m, maximum number of cells; K, carrying capacity; a, constant of logistic kinetic model; r, microalgae growth rate; R^2, correlation coefficient.

Table 1. Experimental data (Cb_o, Nc_o, Cb_m, Nc_m) and logistic model kinetic parameters (K, a, r) determined for the growth of positive controls and treatments with paracetamol of *C. sorokiniana* (CCS+, PCS), *C. vulgaris* (CCV+, PCV) and *S. obliquus* (CSO+, PSO), $n = 3$.

The addition of paracetamol increased the lag phase of the strains of the genus *Chlorella* compared with the positive controls ($p \leq 0.05$), as it can be seen for the values of the parameter a in **Table 1**. However, in the case of *S. obliquus*, the addition of the drug did not modify the beginning of the exponential growth phase compared with the positive control ($p > 0.05$). Furthermore, there were significant differences among the treatments with paracetamol, *C. vulgaris* showed a quite longer lag phase than *S. obliquus* and this one than *C. sorokiniana*.

At the end of the batch culture, the biomass concentration was increased above 49% by the presence of paracetamol in the *C. sorokiniana* culture (CCS+, 1.40 ± 0.29 g l^{-1}; PCS, 2.09 ± 0.02 g l^{-1}) and was increased above 31% in the *C. vulgaris* culture (CCV+, 2.60 ± 0.15 g l^{-1}; PCV, 3.42 ± 0.15 g l^{-1}) compared with their positive control, as shown in **Figure 1** and confirmed by K values in **Table 1**. However, *S. obliquus* culture was not significantly modified by the addition of the drug and the maximum algal density reached in the treatment (PSO, 3.27 ± 0.15 g l^{-1}) was similar to the positive control (CSO+, 3.46 ± 0.08 g l^{-1}). In spite of the different response of the strains to the presence of paracetamol, there were not significant differences between PCV and PSO, even though the value reached for the parameter K in the case of CSO+ was significantly larger than for CCV+. Still, the carrying capacity of the PCS treatment was significantly lower than for PCV and PSO.

Respect to microalgae growth rate (r), there was significant differences between the paracetamol treatment for *C. vulgaris* (PCV, 1.08 ± 0.04 d^{-1}) and the corresponding positive control (CCV+, 0.84 ± 0.06 d^{-1}). However, likewise the K parameter, the growth rate was neither modified in the case of *S. obliquus* treatment (PSO, 1.12 ± 0.03 d^{-1}) compared with the corresponding positive control (CSO+, 1.16 ± 0.07 d^{-1}). Also, no significant differences were detected in the case of *C. sorokiniana* (CCS+, 0.94 ± 0.06 d^{-1}; PCS, 0.96 ± 0.07 d^{-1}). In addition, when comparing the treatments of the three strains, there were not significant differences between the paracetamol treatments of *C. vulgaris* and *S. obliquus* strains (PCV, PSO) despite there were significant differences between their respective positive controls (CCV+, CSO+).

3.1.2. Growth of the culture under salicylic acid addition

The microalgae growth curves of *C. sorokiniana*, *C. vulgaris* and *S. obliquus* under the presence of salicylic acid, the corresponding positive controls and their fittings to the logistic kinetic model, are represented as values of biomass concentration versus time in **Figure 2**. The differences among the treatments were analysed according to growth kinetic parameters, as shown in **Table 2**.

	CCS+	SCS	CCV+	SCV	CSO+	SSO
Cb_0 (g l^{-1})	0.04	0.04	0.11	0.11	0.08	0.08
Nc_0 (cell ml^{-1})	3.20 ×10^6	3.20 ×10^6	1.21×10^6	1.21×10^6	8.35×10^5	8.35×10^5
Cb_m (g l^{-1})	1.41 ± 0.29	2.05 ± 0.15	2.48 ± 0.11	3.02 ± 0.27	3.33 ± 0.06	4.33 ± 0.30
Nc_m (cell ml^{-1})	2.12 × 10^8 ± 0.49 × 10^8	3.15 × 10^8 ±0 0.08 × 10^8	1.18 × 10^8 ± 0.09 × 10^8	1.76 × 10^8 ± 0.49 × 10^8	4.77 × 10^7 ± 0.01 × 10^7	6.97 × 10^7 ± 0.20 × 10^7
a	3.77 ± 0.01	4.16 ± 0.48	4.47 ± 0.25	7.99 ± 0.41	5.45 ± 0.43	4.20 ± 0.09
K (g l^{-1})	1.40 ± 0.29	2.14 ± 0.13	2.60 ± 0.15	3.09 ± 0.23	3.46 ± 0.08	4.71 ± 0.30
r (d^{-1})	0.94 ± 0.06	0.77 ± 0.12	0.84 ± 0.06	1.69 ± 0.13	1.16 ± 0.07	0.72 ± 0.01
R^2	0.9935	0.9912	0.9971	0.9929	0.9874	0.9883

Cb_o, initial biomass; Nc_o, initial number of cells; Cb_m, maximum biomass; Nc_m, maximum number of cells; *K*, carrying capacity; *a*, constant of logistic kinetic model; *r*, microalgae growth rate; R^2, correlation coefficient.

Table 2. Experimental data (Cb_o, Nc_o, Cb_m, Nc_m) and logistic model kinetic parameters (*K*, *a*, *r*) determined for the growth of positive controls and treatments with salicylic acid of *C. sorokiniana* (CCS+, SCS), *C. vulgaris* (CCV+, SCV) and *S. obliquus* (CSO+, SSO), *n*=3 .

Regarding the parameter *a*, the addition of salicylic acid increased significantly the lag phase of the strains *C. vulgaris* and *S. obliquus* compared with the positive controls. Also, *C. sorokiniana* treatment showed a higher *a* value than the positive control, in spite of the difference being not significant (**Table 2**). Comparing the treatments with salicylic acid, *C. vulgaris* showed a quite longer lag phase than *C. sorokiniana* and *S. obliquus*.

As it can be seen in **Figure 2**, the maximum algal density reached at the end of the batch culture was significantly higher in the treatments with salicylic acid for all strains here considered as compared with the positive controls. The *C. sorokiniana* treatment increased their biomass concentration above 52% (CCS+, 1.40 ± 0.29 g l^{-1}; SCS, 2.14 ± 0.13 g l^{-1}), *C. vulgaris* above 18% (CCV+, 2.60 ± 0.15 g l^{-1}; SCS, 3.09 ± 0.23 g l^{-1}) and *S. obliquus* above 36% (CSO+, 3.46 ± 0.08 g l^{-1}; SCS, 4.71 ± 0.30 g l^{-1}) over their respective positive controls at the end of the batch culture. However, under salicylic acid, the carrying capacity of *S. obliquus* was significantly larger than those of *C. sorokiniana* and *C. vulgaris*.

The *C. vulgaris* growth rate was significantly increased under the presence of salicylic acid in comparison with the positive control (CCV+, 0.84 ± 0.06 d^{-1}; SCV, 1.69 ± 0.13 d^{-1}). However, it was significantly reduced in the case of *C. sorokiniana* (CCS+, 0.94 ± 0.06 d^{-1}; SCS, 0.77 ± 0.12 d^{-1}) and *S. obliquus* (CSO+ 1.16 ± 0.07 d^{-1}, SSO, 0.72 ± 0.01 d^{-1}). Moreover, the growth rate of SSO was significantly lower than that of SCS and SCV.

3.1.3. Growth of the culture under diclofenac addition

The microalgae growth curves of *C. sorokiniana*, *C. vulgaris* and *S. obliquus* under the presence of diclofenac, the corresponding positive controls and their respective fittings to the logistic kinetic model are represented as values of biomass concentration versus time in **Figure 3**. The differences among the treatments were analysed according to growth kinetic parameters, as shown in **Table 3**.

	CCS+	DCS	CCV+	DCV	CSO+	DSO
Cb_0 (g l^{-1})	0.04	0.04	0.23	0.23	0.14	0.14
Nc_0 (cell ml^{-1})	3.39×10^6	3.39×10^6	3.53×10^6	3.53×10^6	3.40×10^6	3.40×10^6
Cb_m (g l^{-1})	1.53 ± 0.11	2.28 ± 0.03	1.69 ± 0.06	2.51 ± 0.13	1.27 ± 0.04	1.40 ± 0.05
Nc_m (cell ml^{-1})	2.49 × 10^8 ± 0.22 × 10^8	4.19 × 10^8 ± 0.04 × 10^8	7.91 × 10^7 ± 0.19 × 10^7	1.73 × 10^8 ± 0.22 × 10^8	5.15 × 10^7 ± 0.38 × 10^7	6.33 × 10^7 ± 0.32 × 10^7
a	3.31 ± 0.16	4.24 ± 0.00	2.60 ± 0.05	3.57 ± 0.12	3.30 ± 0.24	3.76 ± 0.37
K (g l^{-1})	1.58 ± 0.11	2.30 ± 0.03	1.96 ± 0.13	2.65 ± 0.10	1.34 ± 0.03	1.49 ± 0.05
r (d^{-1})	0.72 ± 0.04	0.96 ± 0.01	0.56 ± 0.00	0.74 ± 0.01	0.79 ± 0.03	0.81 ± 0.09
R^2	0.9907	0.9988	0.9804	0.9915	0.9890	0.9860

Cb_o, initial biomass; Nc_o, initial number of cells; Cb_m, maximum biomass; Nc_m, maximum number of cells; *K*, carrying capacity; *a*, constant of logistic kinetic model; *r*, microalgae growth rate; R^2, correlation coefficient.

Table 3. Experimental data (Cb_o, Nc_o, Cb_m, Nc_m) and logistic model kinetic parameters (*K, a, r*) determined for the growth of positive controls and treatments with diclofenac of *C. sorokiniana* (CCS+, DCS), *C. vulgaris* (CCV+, DCV) and *S. obliquus* (CSO+, DSO), $n = 3$.

There were significant differences respect the parameter *a* ($p \leq 0.05$) between the positive control and the corresponding treatment of each strain of microalgae, reaching higher values in the case of the treatments with diclofenac. Therefore, the presence of diclofenac produced a delayed response in the beginning of the exponential growth phase compared with the positive control. Comparing the treatments with diclofenac, *C. sorokiniana* showed a longer lag phase than *C. vulgaris* and *S. obliquus*

As it can be seen in **Figure 2**, the treatments with diclofenac achieved significantly higher biomass concentration than their respective positive controls. At the end of the batch culture, the *C. sorokiniana* treatment showed an increase of biomass concentration above 45% (CCS+, 1.58 ± 0.11 g l^{-1}; DCS, 2.30 ± 0.03 g l^{-1}), *C. vulgaris* above 35% (CCV+, 1.96 ± 0.13 g l^{-1}; SCV, 2.65 ± 0.10 g l^{-1}) and *S. obliquus* above 11% (CSO+, 1.34 ± 0.03 g l^{-1}; SCS, 1.49 ± 0.05 g l^{-1}) over their respective positive controls. The *C. vulgaris* treatment reached the highest

K value, which was significantly higher than those determined for the *C. sorokiniana* and the *S. obliquus* treatments.

With respect to microalgae growth rate, there were significant differences between the positive control and the corresponding treatment for the two strains of the genus *Chlorella* here used. The *C. sorokiniana* growth rate was significantly increased under the presence of this drug (CCV+, 0.72 ± 0.04 d^{-1}; DCS, 0.96 ± 0.01 d^{-1}). This significant increase was also confirmed for *C. vulgaris* (CCV+, 0.56 ± 0.00 d^{-1}, DCV, 0.74 ± 0.01 d^{-1}). However, no significant differences were determined in the case of *S. obliquus* (CSO+, 0.79 ± 0.03 d^{-1}, DSO, 0.81 ± 0.09 d^{-1}).

3.2. Removal of pharmaceuticals

The pharmaceutical concentration in each reactor was daily monitored and compared with the concentration of each pharmaceutical in the corresponding negative control. The concentration of the pharmaceuticals here studied decreased over the time in the treatments with microalgae, either with *C. sorokiniana* (PCS, SCS, DCS), *C. vulgaris* (PCV, SCV, DCS) or *S. obliquus* (PSO, SSO, DSO). Meanwhile, no concentration reduction was observed in the negative controls (CP−, CS−, CD−). Therefore, it may be assumed that the pharmaceuticals concentration decrease in the microalgae treatments was due to the removal by the microalgae.

3.2.1. *Removal of paracetamol*

The removal curves of paracetamol by each strain of microalgae and the corresponding fittings to the logistic kinetic model during the batch mode are displayed in **Figure 4(a)**. In addition, differences among the treatments were analysed according to removal kinetic parameters, as shown in **Table 4**.

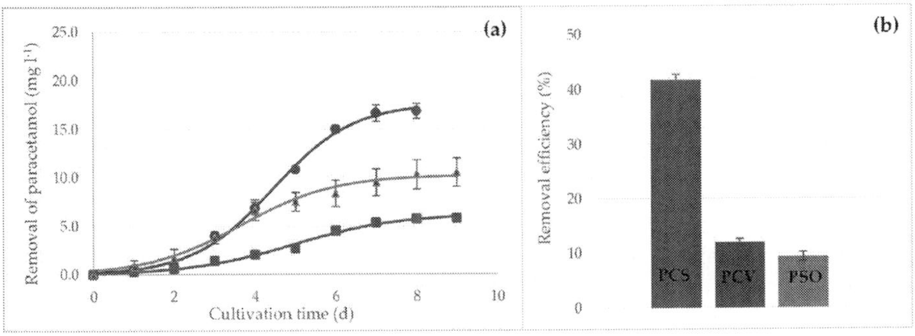

Figure 4. Volumetric efficiency in the removal of paracetamol by *C. sorokiniana* (PCS, •), *C. vulgaris* (PCV, ▪) and *S. obliquus* (PSO, ▴) during batch culture (a). Dots correspond to experimental data and continuous lines correspond to fittings by the logistic kinetic model during batch culture. Volumetric efficiency in the removal of paracetamol (b) at the steady state of the semicontinuous culture. Experiments were performed in triplicate and bars show standard derivations.

Regarding the parameter *a*, there were no significant differences between *C. vulgaris* and *S. obliquus* in the lag phase for the removal of paracetamol. However, *C. sorokiniana* showed a significantly longer response at the beginning of the removal of this drug than the other two strains.

Paracetamol	PCS	PCV	PSO
a	4.49 ± 0.24	3.84 ± 0.01	3.19 ± 0.58
K (mg l^{-1})	17.62 ± 0.91	6.23 ± 0.02	10.41 ± 1.58
r (d^{-1})	1.01 ± 0.06	0.77 ± 0.01	0.86 ± 0.21
R^2	0.9941	0.9827	0.9766
Volumetric efficiency (mg l^{-1} d^{-1})	3.13 ± 0.22	0.95 ± 0.05	0.72 ± 0.07
Specific efficiency (mg g biomass^{-1} d^{-1})	2.68 ± 0.26	0.32 ± 0.02	0.37 ± 0.03
Salicylic acid	**SCS**	**SCV**	**SSO**
a	10.20 ± 3.16	4.09 ± 0.87	4.11 ± 0.16
K (mg l^{-1})	17.68 ± 0.96	6.44 ± 0.63	24.67 ± 0.32
r (d^{-1})	4.07 ± 1.21	0.84 ± 0.17	0.76 ± 0.03
R^2	0.9919	0.9947	0.9973
Volumetric efficiency (mg l^{-1} d^{-1})	6.98 ± 0.31	1.72 ± 0.15	7.55 ± 0.01
Specific efficiency (mg g biomass^{-1} d^{-1})	8.34 ± 1.21	0.67 ± 0.06	1.85 ± 0.02
Diclofenac	**DCS**	**DCV**	**DSO**
a	3.88 ± 0.62	3.23 ± 0.02	3.01 ± 0.38
K (mg l^{-1})	14.55 ± 0.73	15.52 ± 0.26	22.43 ± 0.20
r (d^{-1})	2.03 ± 0.33	1.44 ± 0.05	1.25 ± 0.19
R^2	0.9626	0.9755	0.9690
Volumetric efficiency (mg l^{-1} d^{-1})	2.18 ± 0.39	1.53 ± 0.32	5.66 ± 0.39
Specific efficiency (mg g biomass^{-1} d^{-1})	1.73 ± 0.38	0.97 ± 0.19	5.21 ± 0.18

Table 4. Logistic model kinetic parameters (*K, a, r*) determined for the removal of paracetamol, salicylic acid and diclofenac in the batch culture of *C. sorokiniana* (PCS, SCS, DCS), *C. vulgaris* (PCV, SCV, DCV) and *S. obliquus* (PSO, SSO, DSO). Volumetric efficiency and specific efficiency attained in the steady state of the semicontinuous culture. n=3.

The parameter *K* values in **Table 4** revealed that *C. sorokiniana* (PCS, 17.62 ± 0.91 mg l^{-1}) reached a carrying capacity 2.8 times higher than *C. vulgaris* (PCV, 6.23 ± 0.02 mg l^{-1}) and 1.7 times higher than *S. obliquus* (PSO, 10.41 ± 1.58 mg l^{-1}). In the same way, the removal rates revealed significant differences among the treatments, with *C. sorokiniana* showing the quickest removal (PCS, 1.01 ± 0.06 d^{-1}) and *C. vulgaris* the slowest one (PCV, 0.77 ± 0.01 d^{-1}), which is in agreement with the determined *K* values.

As a consequence of the different responses obtained for the removal parameters between the strains, at the end of the batch culture, efficiencies in the removal of paracetamol above 67% for *C. sorokiniana*, 21% for *C. vulgaris* and 40% for *S. obliquus* were achieved. These results evidenced a larger removal capacity of paracetamol by *C. sorokiniana*, followed by *S. obliquus*, and *C. vulgaris*.

The average volumetric efficiencies on the paracetamol removal by each strain at the steady stage of the semicontinuous culture are depicted as percentages in **Figure 4(b)**. The paracetamol volumetric efficiency reached values above 41% for *C. sorokiniana,* 12% for *C. vulgaris* and 9% for *S. obliquus*. Moreover, the ratios between the volumetric efficiency and the microalgae biomass are shown in **Table 4** as specific efficiencies. These results revealed that *C. sorokiniana* cells removed above 7.2 times more paracetamol than *C. vulgaris* and 8.4 times more than *S. obliquus* per gram of biomass. On the other hand, the paracetamol removal per gram of biomass was similar between *S. obliquus* and *C. vulgaris*.

3.2.2. *Removal of salicylic acid*

The removal curves of salicylic acid by each strain of microalgae and the corresponding fittings to the logistic kinetic model during the batch mode are displayed in **Figure 5(a)**. In addition, differences among the treatments were analysed according to removal kinetic parameters, as shown in **Table 4**.

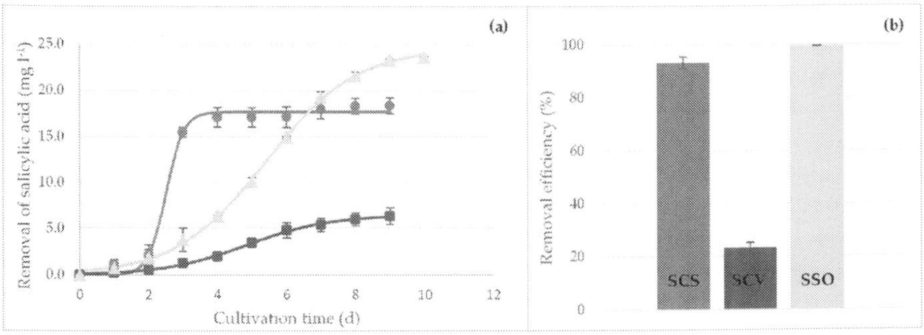

Figure 5. Volumetric efficiency in the removal of salicylic acid by *C. sorokiniana* (SCS, •), *C. vulgaris* (SCV, ▪) and *S. obliquus* (SSO, ▴) during batch culture (a). Dots correspond to experimental data and continuous lines correspond to fittings by the logistic kinetic model during batch culture. Volumetric efficiency in the removal of salicylic acid (b) at the steady state of the semicontinuous culture. Experiments were performed in triplicate and bars show standard derivations.

In the case of *C. sorokiniana* there were significant differences respect the parameter *a,* which indicated that the beginning of the removal of salicylic acid had a delayed response as compared with the lag phase of *C. vulgaris* and *S. obliquus*.

The results obtained for the maximum removal capacity (*K* parameter) revealed that *S. obliquus* (SSO, 24.67 ± 0.32 mg l^{-1}) removed 1.4 times more salicylic acid than *C. sorokiniana* (SCS, 17.68 ± 0.96 mg l^{-1}) and 3.8 time more than *C. vulgaris* (SCV, 6.44 ± 0.63 mg l^{-1}). In spite of salicylic acid removal efficiencies at the end of the batch culture being above 73% by *C. sorokiniana,* 25% by *C. vulgaris,* 93% by *S. obliquus,* the removal rate of *S. obliquus* was significantly lower (SSO, 0.76 ± 0.03 d^{-1}) than that of *C. sorokiniana* (SCS, 4.07 ± 1.21 d^{-1}).

The average salicylic acid volumetric efficiencies by each strain at the steady stage of the semicontinuous culture are depicted as percentages in **Figure 5(b)**. The paracetamol volumetric efficiency did not showed significant differences between the strains *C. sorokiniana* and *S. obliquus*, reaching values above 93% for SCS and 99% for SSO. However, the salicylic acid volumetric efficiency of *C. vulgaris* (above 22%) was more than four times lower than by the other strains. Moreover, the obtained specific efficiencies revealed that *C. sorokiniana* removed above 12.4 times more salicylic acid than *C. vulgaris* and 4.5 times more than *S. obliquus* per gram of biomass (**Table 4**).

3.2.3. *Removal of diclofenac*

The removal curves of diclofenac by each strain of microalgae and the corresponding fittings to the logistic kinetic model during the batch mode are displayed in **Figure 6(a)**. In addition, differences among the treatments were analysed according to removal kinetic parameters, as shown in **Table 4**.

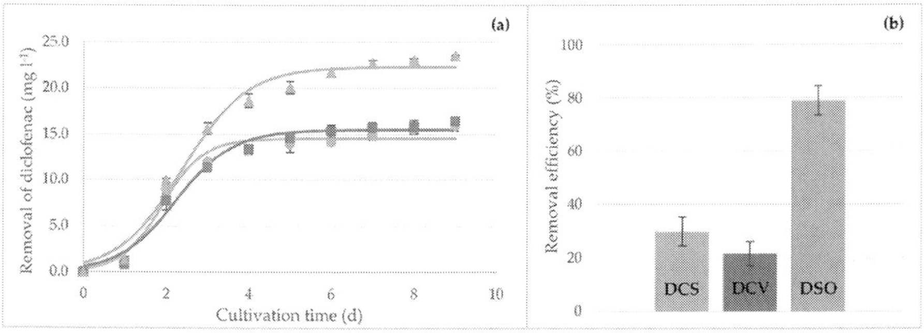

Figure 6. Volumetric efficiency in the removal of diclofenac by *C. sorokiniana* (DCS, •), *C. vulgaris* (DCV, ▪) and *S. obliquus* (DSO, ▴) during batch culture (a). Dots correspond to experimental data and continuous lines correspond to fittings by the logistic kinetic model during batch culture. Volumetric efficiency in the removal of salicylic acid (b) at the steady state of the semicontinuous culture. Experiments were performed in triplicate and bars show standard derivations.

The *a* values were similar ($p > 0.05$) for all the treatments, which indicated that the three strains showed the same delayed response in the removal of diclofenac. However, regarding the maximum removal capacity, there were significant differences between the treatment with *S. obliquus* (DSO, 22.43 ± 0.20 mg l^{-1}), which removed 1.5 times more diclofenac than by *C. sorokiniana* (DCS, 14.55 ± 0.73 mg l^{-1}) and 1.4 times more than *C. vulgaris* (DSO, 15.52 ± 0.26 mg l^{-1}).

Concerning the removal rate, the obtained results revealed significant differences among the treatments. The quickest removal rate was attained by *C. sorokiniana* (DCS, 2.03 ± 0.33 d^{-1}), with removal values 1.6 times higher than *S. obliquus* (DSO, 1.25 ± 0.19 d^{-1}) and 1.4 times higher than *C. vulgaris* (DCS, 1.44 ± 0.05 d^{-1}). Despite the differences between strains regarding the removal parameters, at the end of the batch culture, efficiencies above 65% for *C. sorokiniana*, 69% for *C. vulgaris* and 98% for *S. obliquus* were achieved.

The average volumetric efficiencies for the diclofenac removal in the steady stage of the semicontinuous culture are showed in **Figure 6(b)**. The volumetric efficiency for *S. obliquus* (above 79%) was 2.6 times higher than for *C. sorokiniana* and 3.7 times higher than for *C. vulgaris*.

http://dx.doi.org/10.5772/66772

Moreover, the ratios between the volumetric efficiency and the microalgae biomass are shown in **Table 4** as specific efficiencies. The determined values revealed that *S. obliquus* removed above 3.0 times more diclofenac than *C. sorokiniana* and above 5.4 times more than *C. vulgaris* per gram of biomass.

4. Discussion

In view of the obtained results, it may be inferred that the presence of paracetamol, salicylic acid and diclofenac modified the growth parameters of the strains here studied. In most of the treatments, the addition of the pharmaceutical increased the biomass concentration, which may be explained by the fact that these pharmaceuticals were an additional source of organic carbon. It is well known that the genus *Chlorella* and *Scenedesmus* can have a mixotrophic growth. However, *S. obliquus* did not show a significant increase of microalgae biomass under the addition of paracetamol or diclofenac. These results suggest that the other removal mechanisms, apart from metabolism, may be involved.

The fact that removal curves displayed a similar trend than growth curves points to the association between the microalgae growth and the removal efficiency of pharmaceuticals.

In view of the obtained results, it may be concluded that paracetamol was more efficiently removed by *C. sorokiniana*, either per litre or per gram of biomass (>67% batch culture, >41% semicontinuous culture), in spite of the biomass concentration reached in the culture being the lowest one among the three strains. Also, the removal rate by *C. sorokiniana* was the fastest one, in spite of showing the lowest growth rate among the paracetamol treatments. However, the addition of paracetamol in the *C. sorokiniana* culture produced the largest increase in the biomass concentration compared with the corresponding positive control (>49%).

On the other hand, *S. obliquus* showed the highest salicylic acid removal capacity at the end of the batch culture (>93%) and also at the steady state of the semicontinuous culture (>99%). However, the removal rate by *S. obliquus* was the lowest one among the salicylic acid treatments. The highest removal rate was reached by *C. sorokiniana*, which showed a removal per gram of biomass 4.5 times larger than *S. obliquus*. Furthermore, the increase of biomass under the addition of salicylic acid was above 52% in the *C. sorokiniana* treatment, while for *S. obliquus* was above 36%.

Regarding diclofenac, despite *C. sorokiniana* cells attained a higher removal rate and the higher growth rate, it may be stated that *S. obliquus* was the strain that reached the highest removal efficiency (>98% batch culture, >79% semicontinuous culture) with more diclofenac removed either per litre or per gram of biomass.

Comparing the three pharmaceuticals, the salicylic acid was more efficiently removed, with *C. sorokiniana* and *S. obliquus* showing the highest efficiencies. Contrarily, the paracetamol was the less efficiently removed. In all cases, *C. vulgaris* showed the lowest efficiencies for the three pharmaceuticals. These results may be related with the specific strain characteristics, the mechanisms involved in the removal and the particular properties of each pharmaceutical.

As in this work, published results on the removal of ECs by microalgae have revealed different efficiencies depending on the pollutant and on the microalgae strain. For example, Gattullo et al. [23] demonstrated that *Monoraphidium braunii* was able to remove up to 48% of bisphenol A with an initial concentration of 4 mg l^{-1}. de Wilt et al. [14] reported removal efficiencies by *C. sorokiniana*, grown in wastewater streams, up to 60–100% for diclofenac, ibuprofen, paracetamol and metoprolol. However, under identical conditions, the removal of carbamazepine and trimethoprim was incomplete and did not exceed 30% and 60%, respectively [14]. Wang et al. [24] studied the removal of phenol by *Chorella* sp. culture, obtaining removal efficiencies up to 100% from an initial concentration of 500 mg l^{-1} in 7 days. Peng et al. [25] reported removals above 95% of progesterone by *S. obliquus* and *Chlorella pyrenoidosa*, nearly complete removal of norgestrel by *S. obliquus* and almost 40% of norgestrel by *C. pyrenoidosa*. Likewise, Hom-Díaz et al. [15] studied the elimination of the hormones E2 and EE2 from anaerobic digestate centrate by the microalgae *Selenastrum capricornutum* and *Chlamydomonas reinhardtii*. After 7 days of culture, these authors [15] determined removals above 88% for E2 and above 60% for EE2. Furthermore, Matamoros et al. [26] studied the capability of microalgae-based wastewater treatment systems to remove diclofenac, among other 25 emerging organic contaminants. These authors [26] determined diclofenac removal efficiencies above 82% under HRT of 4 days and above 92% under HRT of 8 days during the warm season (11–26°C, on a daily average). These efficiencies are higher than the here obtained under an HRT of 80 h and temperature of 25±1°C. Differences must be related, at least to some extent, to the fact that microalgae monocultures were used in this work while Matamoros et al. [26] worked with mixed microalgae strains present in the wastewater, mostly identified as *Stigeoclonium* sp., diatoms, *Chlorella* sp. and *Monoraphidium* sp.

5. Conclusions

Among the here considered strains, *S. obliquus* displayed the highest removal efficiency for salicylic acid and diclofenac, while *C. sorokiniana* did it for paracetamol. On the other hand, *C. vulgaris* showed the lowest efficiencies for the three pharmaceuticals. Comparing the three pharmaceuticals, the salicylic acid was more efficiently removed while paracetamol removal was the less efficient. These differences may be related with the specific strain characteristics, the mechanisms involved in the removal and the particular properties of each pharmaceutical. The obtained results pointed to the feasibility of using the microalgae here considered in bioremediation systems and revealed that this sort of studies are key for the selection of the strain, which depends on the application. Still, further research is needed to assess the mechanisms involved in the removal of pharmaceuticals by these strains.

Acknowledgements

Authors thank University of León for funding given to MICROTRAT (project UXXI2016/00128). Carla Escapa and Sergio Paniagua acknowledge the Spanish Ministry of Educations, Culture and Sports for their PhD fellowships (FPU12/03073 and FPU14/05846, respectively). Marta Otero acknowledges University of León for the extension of her RYC-2010-05634 contract.

Author details

Carla Escapa Santos, Ricardo Nuno de Coimbra, Sergio Paniagua Bermejo, Ana Isabel García Pérez* and Marta Otero Cabero*

*Address all correspondence to: ana.garcia@unileon.es and marta.otero@unileon.es

Department of Applied Chemistry and Physics, Institute of Environment, Natural Resources and Biodiversity (IMARENABIO), University of León, León, Spain

References

[1] Sauvé S, Desrosiers M: A review of what is an emerging contaminant. Chemistry Central Journal. 2014;**8**:15. DOI: 10.1186/1752-153X-8-15

[2] Evgenidou E, Konstantinou I, Lambropoulou D: Occurrence and removal of transformation products of PPCPs and illicit drugs in wastewaters: A review. Science of the Total Environment. 2015;**505**:905–926. DOI: 10.1016/j.scitotenv.2014.10.021

[3] Santos L, Araújo A, Fachini A, Pena A, Delerue-Matos C, Montenegro M: Ecotoxicological aspects related to the presence of pharmaceuticals in the aquatic environment. Journal of Hazardous Materials. 2010;**175**:45–95. DOI: 10.1016/j.jhazmat.2009.10.100

[4] Barbosa M, Moreira N, Ribeiro A, Pereira M, Silva A: Occurrence and removal of organic micropollutants: An overview of the watch list of EU Decision 2015/495. Water Research. 2016;**94**:257–279. DOI: 10.1016/j.watres.2016.02.047

[5] Rigobello E, Dantas A, Di Bernardo L, Vieira E: Removal of diclofenac by conventional drinking water treatment processes and granular activated carbon filtration. Chemosphere. 2013;**93**:184–191. DOI: 10.1016/j.chemosphere.2013.03.010

[6] Barceló D, Petrovic M: Reducing the environmental risk from emerging pollutants. In: Trac-Trends in Analytical Chemistry, editor. 1st EMCO workshop "Analysis and removal of contaminants from wastewater for the implementation of the Water Framework Directive (WFD)"; 20-21 October 2005; Dubrovnik, Croatia. 2006. p. 191–193. DOI: 10.1016/j.trac.2005.11.003

[7] Bolong N, Ismail A, Salim M, Matsuura T: A review of the effects of emerging contaminants in wastewater and options for their removal. Desalination. 2009;**239**:229–246. DOI: 10.1016/j.desal.2008.03.020

[8] Farré M, Pérez L, Kantiani L, Barceló D: Fate and toxicity of emerging pollutants, their metabolites and transformation products in the aquatic environment. Trac-Trends in Analytical Chemistry. 2008;**27**:991–1007. DOI: 10.1016/j.trac.2008.09.010

[9] Pal A, He Y, Jekel M, Reinhard M, Gin K: Emerging contaminants of public health significance as water quality indicator compounds in the urban water cycle. Environment International. 2014;**71**:46–62. DOI: 10.1016/j.envint.2014.05.025

[10] Verlicchi P, Aukidy M, Zambello E: Occurrence of pharmaceutical compounds in urban wastewater: Removal, mass load and environmental risk after a secondary treatment-A review. Science of the Total Environment. 2012;**429**:123–155. DOI: 10.1016/j.scitotenv.2012.04.028

[11] Heberer T, Reddersen K, Mechlinski A: From municipal sewage to drinking water: Fate and removal of pharmaceutical residues in the aquatic environment in urban areas. Water Science and Technology. 2002;**46**:81–88.

[12] Ashton D, Hilton M, Thomas K: Investigating the environmental transport of human pharmaceuticals to streams in the United Kingdom. Science of the Total Environment. 2004;**333**:167–184. DOI: 10.1016/j.scitotenv.2004.04.062

[13] Combarros R, Rosas I, Lavin A, Rendueles M, Diaz M: Influence of biofilm on activated carbon on the adsorption and biodegradation of salicylic acid in wastewater. Water, Air, and Soil Pollution. 2014;**225**:1858. DOI: 10.1007/s11270-013-1858-9

[14] de Wilt A, Butkovskyi A, Tuantet K, Leal L, Fernandes T, Langenhoff A, Zeeman G: Micropollutant removal in an algal treatment system fed with source separated wastewater streams. Journal of Hazardous Materials. 2016;**304**:84–92. DOI: 10.1016/j.jhazmat.2015.10.033

[15] Hom-Díaz A, LLorca M, Rodríguez-Mozaz S, Vicent T, Barceló D, Blanquez P: Microalgae cultivation on wastewater digestate: Beta-estradiol and 17 alpha-ethynylestradiol degradation and transformation products identification. Journal of Environmental Management. 2015;**155**:106–113. DOI: 10.1016/j.jenvman.2015.03.003

[16] Ritchmon A: Handbook of Microalgal Mass Culture. Boca Raton, Florida: CRC Press; 1986. 528 p.

[17] Matamoros V, Uggetti E, García J, Bayona J: Assessment of the mechanisms involved in the removal of emerging contaminants by microalgae from wastewater: A laboratory scale study. Journal of Hazardous Materials. 2016;**301**:197–205. DOI: 10.1016/j.jhazmat.2015.08.050

[18] Petrie B, Barden R, Kasprzyk-Hordern B: A review on emerging contaminants in wastewaters and the environment: Current knowledge, understudied areas and recommendations for future monitoring. Water Research. 2015;**72**:3–27. DOI: 10.1016/j.watres.2014.08.053

[19] Beuckels A, Smolders E, Muylaert K: Nitrogen availability influences phosphorus removal in microalgae-based wastewater treatment. Water Research. 2015;**77**:98–106. DOI: 10.1016/j.watres.2015.03.018

[20] Mann J, Myers J: On Pigments Growth and Photosynthesis of *Phaeodactylum tricornutum*. Journal of Phycology. 1968;**4**:349–355. DOI: 10.1111/j.1529-8817.1968.tb04707.x

[21] Verhulst P: *Notice sur la loi que la population suit dans son accroissement* (A note on the law of population growth). Correspondace Mathématique et Physique. 1938;**10**:113–121.

[22] Xin L, Hu H, Ke G, Sun Y: Effects of different nitrogen and phosphorus concentrations on the growth, nutrient uptake, and lipid accumulation of a freshwater microalga *Scenedesmus* sp. Bioresource technology. 2010;**101**:5494–5500. DOI: 10.1016/j.biortech.2010.02.016

[23] Gattullo C, Baehrs H, Steinberg C, Loffredo E: Removal of bisphenol A by the freshwater green alga *Monoraphidium braunii* and the role of natural organic matter. Science of the Total Environment. 2012;**416**:501–506. DOI: 10.1016/j.scitotenv.2011.11.033

[24] Wang L, Xue C, Wang L, Zhao Q, Wei W, Sun Y: Strain improvement of *Chlorella* sp for phenol biodegradation by adaptive laboratory evolution. Bioresource Technology. 2016;**205**:264–268. DOI: 10.1016/j.biortech.2016.01.022

[25] Peng F, Ying G, Yang B, Liu S, Lai H, Liu Y, Chen Z, Zhou G: Biotransformation of progesterone and norgestrel by two freshwater microalgae (*Scenedesmus obliquus* and *Chlorella pyrenoidosa*): Transformation kinetics and products identification. Chemosphere. 2014;**95**:581–588. DOI: 10.1016/j.chemosphere.2013.10.013

[26] Matamoros V, Gutiérrez R, Ferrer I, García J, Bayona J: Capability of microalgae-based wastewater treatment systems to remove emerging organic contaminants: A pilot-scale study. Journal of hazardous materials. Journal of Hazardous Materials. 2015;**288**:34–42. DOI: 10.1016/j.jhazmat.2015.02.002

Chapter 7

Pulp Mill Wastewater: Characteristics and Treatment

María Noel Cabrera

Additional information is available at the end of the chapter

http://dx.doi.org/10.5772/67537

Abstract

The production of chemical pulp in recent times is 180 million tons per year; while the production of eucalyptus pulp has increased intensively, especially in the southern hemisphere. The pulp and paper industry has long been considered a large consumer of natural resources (wood and water) and one of the largest sources of pollution to the environment (air, water courses and soil). Important efforts are being made to reduce the pollutant levels and water consumption of the industry. The wastewater composition, and therefore, the efficiency of effluent treatments and characteristics of the discharges to water are strongly dependent on the applied technology and raw materials. Despite a large body of literature on softwood-based wastewater, few studies have examined the characteristics of kraft eucalyptus bleaching effluents and their behaviour in the different biological treatments. The largest secondary treatment systems today use the activated sludge process. Sixty to seventy-five per cent of all the biological effluent treatment plants within the pulp and paper industry use this kind of treatment system. This chapter reviews the current pulping technologies at mills and compares the chemical composition and biological treatment of wastewater between softwood and hardwood bleached pulps.

Keywords: pulp mills, hardwood, softwood, kraft pulping, ECF-TCF bleaching

1. Introduction

The pulp and paper mill industry is an intensive consumer of water and natural resources (wood), discharging a variety of liquid, gaseous and solid wastes to the environment. Since the 1970s, a growing awareness of the effects of pulp and paper wastes in the ambience had prompted water and energy consumption levels and the loads of toxic compounds discharge to reduce. One of the most important implemented changes in this regard was made within the mill, wherein chlorine was completely substituted by, that is, chlorine dioxide as the bleaching chemical agent. Another major issue was the implementation of secondary biological treatments. The wastewater composition and hence the effluent treatment efficiencies and characteristics of the discharges are strongly dependent on the technology applied and the raw materials. In the last 25 years, however, the global distribution of pulp producers has significantly changed and so have the species of wood used. Eucalyptus pulp production, for example, is becoming a leader in the hardwood pulp market; Brazil went from being a pulp consumer to a world leader in hardwood pulp production, and since 2008, it has been the fourth largest pulp producer in the world.

2. Wood pulp market

Cellulose pulp is the main raw material in the production of different types of paper and paperboard. It is also used as the absorbent material in diapers and other sanitary products.

The global pulp market has changed intensely in recent years. A few decades ago, this industry was characterized as national character as a supply industry inputs for domestic production of paper and paperboard. Globalization has led to increased competitiveness in the international market, as new players have emerged both at the level of producers and consumers. Within the latter, the appearance of China and India have strongly modified cellulose demand worldwide [1].

Figure 1 graphically shows the evolution of world's production of wood pulp between 1979 and 2013 according to the data published by FAO [2–6].

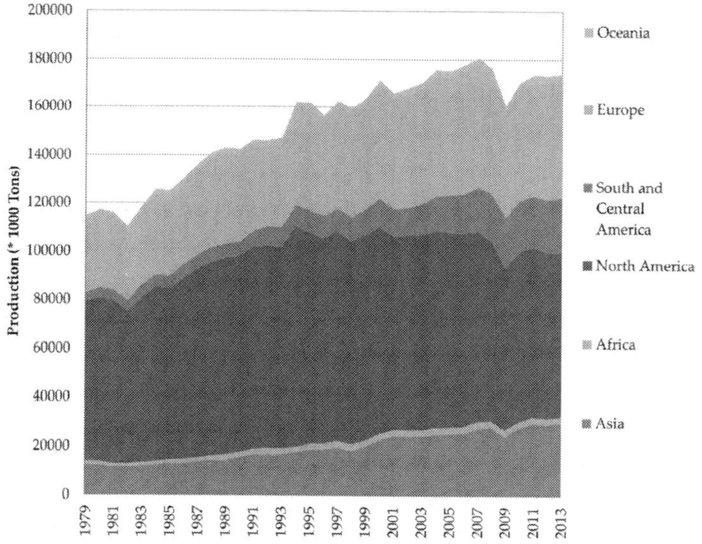

Figure 1. World pulp production 1979–2013. Data obtained from Ref. [2].

http://dx.doi.org/10.5772/67537

It is clearly shown that the world's wood pulp production increased to about 50% in this period, from 120 million tons in 1979 to nearly 180 million tons in 2013. Part of this growth can be explained by the explosive increase in production in non-traditional wood pulp producing regions such as Asia and South America. The main producing regions are still North America with 38% and Europe with 28%, even though in 2013, Asia produced 17% of the wood pulp and South America about 13% [6].

Wood pulp grades are categorized according to the pulping process, which can be classified as mechanical, semi-chemical and chemical pulps. In a mechanical process, logs or wood chips are mechanically grinded by abrasive action. In a chemical cooking process, a significant part of the wood components (mainly lignin) is chemically dissolved to obtain a solid compound with high cellulose fibre content. There are two main methods of chemical pulping: (1) sulphite pulping and (2) sulphate (kraft) pulping. The first process—sulphite cooking process—uses aqueous sulphur dioxide (SO_2) and a base of calcium, sodium, magnesium or ammonium. The kraft process uses a treatment comprising a mixture of sodium hydroxide and sodium sulphide, known as white liquor, at a high pressure and temperature. The semi-chemical pulping process combines chemical and mechanical methods, where wood chips are first softened or partially cooked with chemicals and then mechanically pulped [7].

Figures 2 and **3** illustrate the different kinds of pulp produced in 1979 and 2013.

The rise in wood pulp production is due to an increase in chemical pulp production, as the production of mechanical pulp has declined in the same period. Mechanical pulping has the advantage of converting up to 95% of dry weight wood into pulp, although considerable

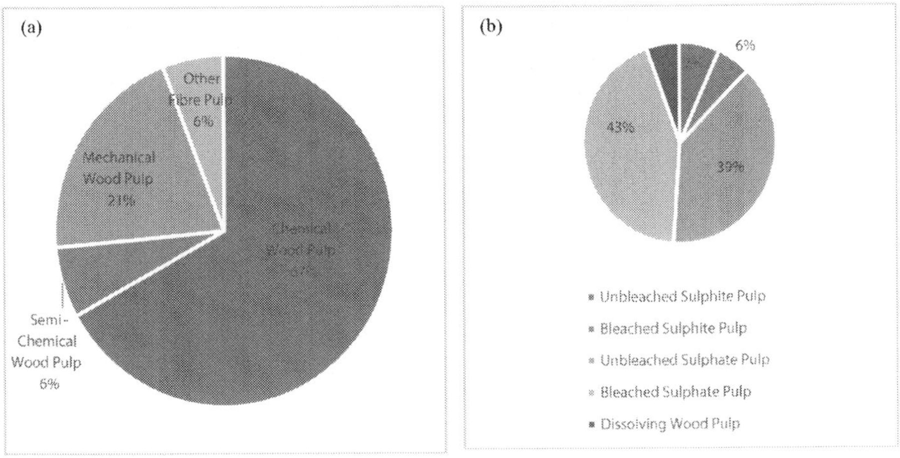

Figure 2. (a) World pulp production by type of pulp in 1979; (b) different kinds of chemical pulps produced in 1979 (Data from FAO [2]).

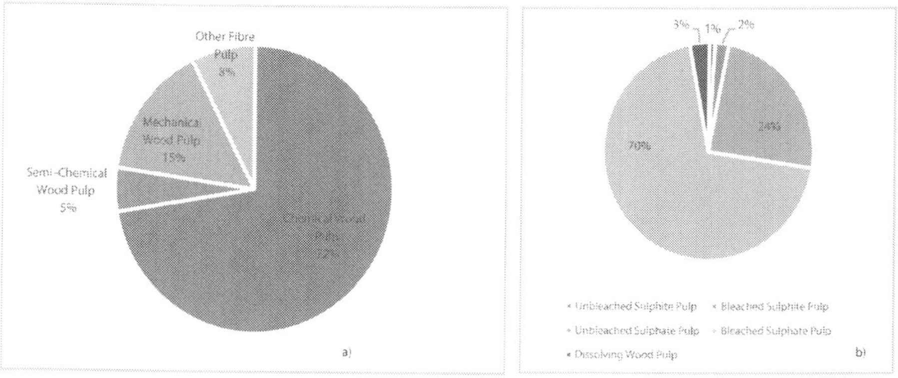

Figure 3. (a) World pulp production by type of pulp in 2013; (b) different kind of chemical pulps produced in 2013 (Data from FAO [6]).

amounts of energy are required to do so. The pulp obtained produces a highly opaque paper with good printability, but the physical properties are inferior than chemical pulps and yellowing when exposed to light. Moreover, mechanical pulps are mainly produced from softwood [7].

There are significant changes in the production of chemical pulp. The use of sulphite cooking process in pulp production compared to kraft pulping technology decreased steadily, from 60% in 1925 to 20% in 1967 and 9.2% in 1979 to only 2.4% in 2013 [6, 8]. The superiority of kraft pulping process is explained by the following facts: (1) all wooden materials including low-quality wood can be used as raw material; (2) superior fibre strength of pulp compared to other chemical pulping methods; (3) more simple chemical and energy recovery process; (4) scale of economy of kraft methods prevents competition and (5) low environmental risks in modern mills [9].

A second classification considers the type of wood used by distinguishing softwood or long fibre (produced mainly from pine and spruce) from hardwood or short fibre (produced from eucalyptus, birch, poplar, etc.) [10]. A gradual move from softwood to hardwood can be observed. In 2013, 56% of bleached kraft pulp was produced with long-fibre wood (softwood), while the remaining 44% was produced with short-fibre wood (hardwood) (according to data from Ref. [6]). In 1980, the production capacity of bleached kraft pulp corresponded to 63% of softwood pulp. The entry into the market of non-traditional producing countries such as Brazil, Indonesia, Spain and Portugal, significantly increased the production of hardwood pulp. Eucalyptus bleached pulp production is rapidly increasing (from 8 million tons in 2003 to nearly 15 million in 2015), and eucalyptus wood is thus considered to be the most important raw material of hardwood bleached market pulp in the world [11].

As kraft pulping is by far the most common process used these days, this chapter will focus in the wastewaters generated in this process.

http://dx.doi.org/10.5772/67537

3. Main processes description

3.1. Mechanical pulping

The oldest method of mechanical pulping is the groundwood process. In this process, round logs are forced against a rotating pulp stone (revolving at peripheral speeds of 1000–1200 m/min), under specified conditions of pressure and temperature. Atmospheric grinding, pressure grinding and thermo-grinding could be done according to the applied temperature and pressure. In all of them, the temperature levels obtained from the heat applied or from rubbing the logs on the stone soften and break down the fibres structure; and cracks the fibres from the wood matrix [7, 8].

Another common method is the refiner mechanical pulping (RMP). The wood chips are pulled between two rotating disks. Among them, thermomechanical pulping operates like RMP, but under higher temperature and pressure. The high temperature and pressure levels soften the lignin even more than frictional heat, making fibres separation easier. Thermomechanical pulp is stronger than refined mechanical pulp, and still retains the high-yield and cost-effectiveness of mechanical pulps [7].

3.2. Chemical pulping

3.2.1. Sulphite pulping

Sulphite process is very versatile, and covers the entire pH range, achieving high fibre flexibility in pulp yields and properties. The cooking process involves the use of aqueous sulphur dioxide (SO_2) and a base: calcium, sodium, magnesium or ammonium. Sulphite pulping was developed in the second half of the nineteenth century and for several decades, the calcium acid sulphite process was the most common method. However, since 1950, the utilization of bases other than calcium has been a major development. The specific base used will determine the process's chemical and energy recovery system and water use. The use of the relatively cheap calcium base has become obsolete because the cooking chemicals cannot be recovered. Magnesium and sodium bases allow chemical recovery, and magnesium bases are currently the dominant choice in sulphite pulping process [7, 12].

3.2.2. Kraft pulping

In kraft pulping, white liquor, containing mainly active chemicals—sodium hydroxide and sodium sulphide—is used for cooking the chips at a high temperature (150–170°C) and pressure. Approximately, half of the wood composition degrades and dissolves during cooking. The spent cooking liquor (black liquor) contains reaction products of lignin and hemicelluloses, and is concentrated and burned in a recovery boiler that recovers the cooking chemicals and generates energy. The smelt is dissolved into water to form green liquor (mostly sodium carbonate and sodium sulphide), which then reacts with lime to convert the sodium carbonate into sodium hydroxide regenerating the white liquor. After cooking and washing, a brown pulp (brown stock pulp) is obtained. Printing, writing and tissue papers require the pulp to be bleached which removes the excess lignin and chromophores to produce a "white" pulp.

4. Background of pulp mill effluents: environmental fate and effects

The pulp and paper industry consumes enormous amounts of water and natural resources and is also one of the largest effluents generators. Before the 1970s, wastewaters from the pulp and paper mills were normally discharged directly to the rivers or lakes, without any treatment or even a rough primary treatment. The high organic loads and solid content in the effluents affected the aquatic ecosystem in several ways such as localized damage to the benthic community, oxygen depletion in large areas and numerous changes in fish reproduction and physiology. In the 1980s, studies in Scandinavia, along the Baltic Coast and the Gulf of Bothnia, showed alterations in fish reproduction and increase of diseases and parasites [13, 14]. Studies conducted in USA and Canada in the beginning of the 1990s, under the Environmental Effects Monitoring (EEM) program [15, 16], revealed delayed sexual maturity, smaller gonads, changes in fish reproduction and depression in secondary sexual characteristics in species living downstream of pulp and paper mills discharges.

From the end of the 1970s until now, the main concern regarding effluents is the formation of chlorinated compounds in bleaching plants. In 1985, 2,3,7,8-tetrachlorodibenzo-p-dioxin (TCDD) was discovered in the pulp mill effluents, which led to a general concern over the formation of chlorinated organic matter in chlorine bleaching. Consequently, the use of chlorine in the bleach plants gradually decreased until it was completely substituted with chlorine dioxide. In many countries, the environmental control authorities set strict restrictions on the discharges of chlorinated organics, measured as adsorbable organic halogen (AOX), into the aquatic environment. In 1992, the Swedish Environmental Protection Agency limited organochlorines emissions to 1.5 kg AOX/t of pulp and in 1995, Finland's official limit was set at 1.4 kg AOX/t of pulp [14].

Several authors reported that with the replacement of chlorine with chlorine dioxide, the effluent quality improved in AOX levels and the elimination of detectable amount of dioxins, polychlorinated compounds and chloroform [12, 13, 17].

The European Integrated Pollution and Prevention Control [12] has created reference documents (BREF) that set the Best Available Techniques (BAT) for several industrial sectors. The pulp and paper industry has a very defined set of operations to be especially applied in the new mills. Similarly, the International Finance Corporation [18] among others has defined directives that could be required to give financial support for the construction of new mills. For kraft pulp, the most important guidelines are listed in **Table 1**.

Dry debarking of wood
Extended modified cooking to a low kappa number (batch or continuous)
Systems for collection and recycling of temporary and accidental discharges from process water spills
Closed screening
Efficient washing of the pulp ahead of the bleaching
Oxygen delignification ahead of the bleach plant
Elemental chlorine free (ECF) or total chlorine free (TCF) bleaching
Removal of hexenuronic acids by mild hydrolysis at the beginning of the bleaching process, for hardwood pulp, especially eucalyptus
Partial closure of the bleach plant combined with increased evaporation
Sufficient and balanced volumes of pulp storage, broke storage and white water storage tanks to avoid or reduce process water discharges
Recycling of wastewater, with or without simultaneous recovery of fibres
Separation of contaminated and non-contaminated (clean) wastewaters
Biological secondary wastewater treatment

Table 1. Main BAT guidelines from IFC [18] and/or IPPC Bureau [12] regarding wastewater load minimization in bleached kraft pulp mills.

5. Mechanical pulping: wastewater characteristics

Figure 4 shows a block diagram of the main part of the mechanical pulp production indicating the sources of emissions to the water from a pulp mill.

Table 2 shows the specific water consumption and loads before wastewater treatment from the mechanical pulping [12].

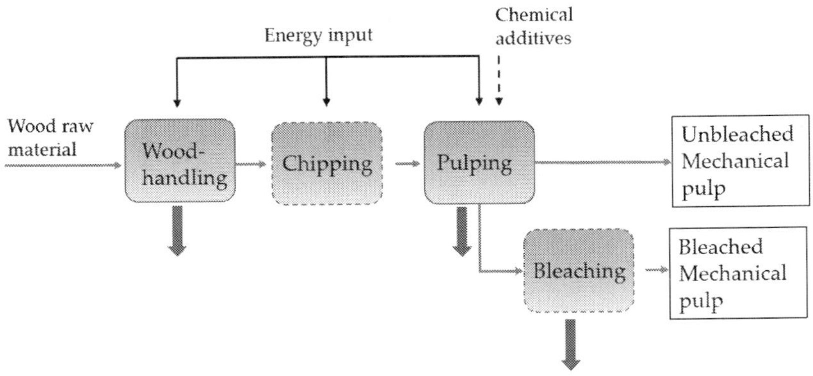

Figure 4. Main unit operations of the mechanical pulping. Light brown arrows indicate wastewater sources.

Pulping process	BOD_5 (kg/ADt)	COD (kg/ADt)	Nitrogen (kg/ADt)	Phosphorous (kg/ADt)
GW	8.5–10	20–30	80–100	20–25
PGW	10–13	30–50	90–110	20–30
RMP	10–15	40–60	90–110	20–30
TMP	13–22	50–80	100–130	30–40

BOD5: Biochemical Oxygen Demand; COD: Chemical Oxygen Demand; GW: groundwood pulping; PGW: pressurized groundwood pulping; RMP: refined mechanical pulping; TMP: thermomechanical pulping; ADt: air dry tone (10% water and 90% oven-dry pulp).

Table 2. Specific water consumption, organic and nutrient loads before wastewater treatment from the mechanical pulping. Data taken from Ref. [12].

6. Kraft pulping: wastewater characteristics

6.1. Process description and emissions to water

A kraft pulp mill can be divided into four main parts: (1) raw material handling; (2) pulping line with an almost closed chemical and energy recovery system; (3) bleaching with an open water system and (4) the external wastewater treatment system. **Figure 5** shows the emissions sources to water from a kraft pulp mill.

Table 3 shows the typical figures for the parameters in different sectors of a kraft pulp mill.

Data on current discharges to water (after wastewater treatment) expressed as loads based on available data from kraft pulp mills within the European Union are given in **Table 4**. **Figure 6** presents a comparison of the discharges to water of different existing mills with the performance of the new mills in South America that are processing eucalyptus wood and apply-

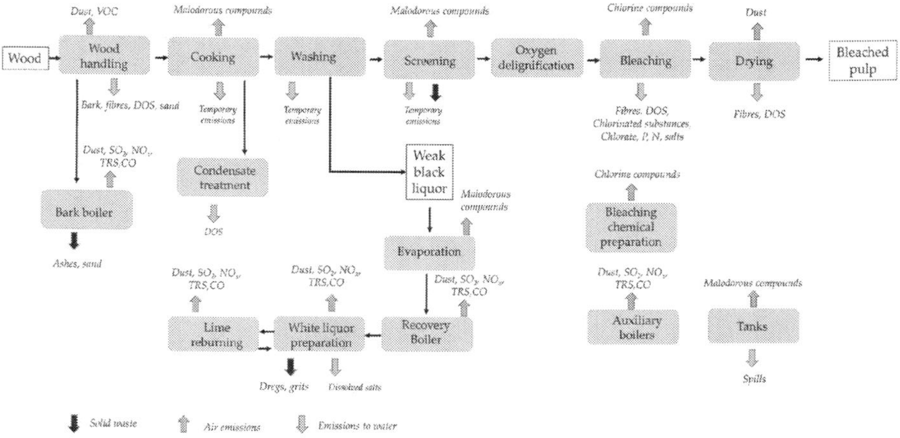

Figure 5. Main unit operations of kraft pulping. DOS: dissolved organic substances. Adapted from Ref. [12].

Department	Flow	TSS	BOD	AOX	COD	P	N
Debarking	2.5	4	2	0	5	20	0. 2
Washing and screening	0.5	3	1	0	2	1	0.015
Bleaching	31	2	10	1.2	35	47	0.075
Condensates	1	0	1	0	3	0	0
Others	3	4	4	0	10	7	0.002
Total	38	13	18	1.2	55	75	300

Flow in m³/Adt, TSS, BOD, AOX, COD and Nitrogen in kg/ADt. Phosphorous in g/ADt.

Table 3. Sources of effluents and effluents loads from kraft pulp mill [12, 19].

	Flow	COD	AOX	TSS	Total P[1]	Total nitrogen
Unbleached pulp	14–82	1.2–23	–	0.02–3	0–0.05	0.01–1.0
Bleached pulp	20–94	5–20[2] 7.5–42[3]	0–0.3	0.015–7	0.003–0.11	0.01–0.6

Flow in m³/ADt, COD, BOD_5, AOX, TSS, nitrogen and phosphorous in kg/ADt.

[1] Eucalyptus strands contain higher levels of phosphorus compared to other forest species used for pulp production. The average level discharged with the effluent is up to 0.12 kg total-P/ADt.

[2] Emissions from eucalyptus pulp mills.

[3] Emissions from other hardwood (no eucalyptus) and softwood.

Table 4. Reported annual average discharges from kraft pulp mills within the EU [12].

ing the Best Available Techniques (according to the European IPPC Bureau [12] and the IFC Guidelines [18]).

6.2. Bleaching effluents

Up to 85% of the total effluent volume is generated in the bleaching stage. Therefore, this part of the mill is broadly studied in order to minimize the effluent organic loads (especially the organochlorines loads) without impacting the pulp yield and brightness. Effluent loadings depend on the production process and the raw materials. The degree of delignification of the unbleached pulp, the bleaching process, the washing loss, type of wood, final brightness desired, chemical and water consumption and the degree of plant closure are important indicators of wastewater characteristics [12, 19]. To this end, kappa number is an important mill control parameter. The kappa number quantifies by a redox reaction to the amount of lignin (or the delignification degree) still in the pulp. The higher the kappa number, the higher the lignin content in the pulp. The low lignin amounts to be removed during bleaching, decreases the utilization of bleaching chemicals, which consequently reduces the load to the wastewater treatment. However, if the kappa number were to decrease too much during the cooking then the pulp yield and physical properties will be considerably low [10]. **Table 5** provides performance data of the different processes [12].

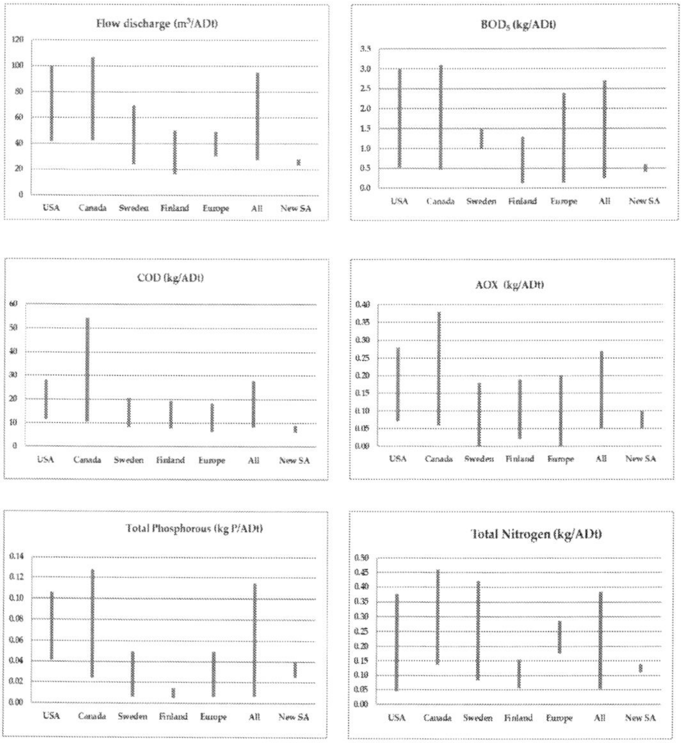

Figure 6. South America new mills performance compared with mills in North America and Europe (Data from EKONO and author personal sources). The vertical bars depicted in the graphs correspond to the 10th percentile to 90th percentile range. The column "All" corresponds to the average of the values reported.

Delignification technologies	Kappa for hardwood	Kappa for softwood	Calculated COD load (kg/t) from the bleach plant	
			Hardwood	Softwood
Conventional cooking	14–22	30–35	28–44	60–70
Conventional cooking + oxygen delignification	13–15	18–20	26–30	36–40
Extended/modified cooking	14–16	18–22	28–32	36–44
Extended cooking + oxygen delignification	8–10	8–12	16–20	16–24

Table 5. Kappa number currently achieved with different delignification technologies and comparison of the calculated effluent COD without considering the washing losses [12].

http://dx.doi.org/10.5772/67537

6.2.1. *Hardwood and softwood bleaching effluents*

The effluents from kraft pulp bleaching constitute varying quantities of organic and inorganic substances. The organic typically represents one-third of the dissolved material while the inorganics comprise two-thirds. The solid matter includes mainly fibres, pieces of fibres and the additives used in bleaching. The dissolved organic matter is composed of various species derived from the raw material and formed in the pulping and bleaching process (residual lignin, hemicelluloses and extractives) [19].

Wood material impact on the values of the effluents parameters can be assessed by comparing the figures for bleaching effluents derived from softwood and hardwood pulp. The former has higher COD and colour content than those of hardwood pulp. The compounds responsible for colour are lignin fragments of high molecular weight (HMW), which represents low biodegradability in the biological treatment [20]. Research has compared effluents from softwood and eucalyptus pulps [13, 20, 21] through AOX, COD, BOD_5 and colour behaviour of the different kinds of pulp production (conventional bleached pulps and oxygen delignified bleached pulps). According to the findings, softwood and eucalyptus effluents have the same trend in AOX levels. For both conventional pulps, the AOX levels were higher than the corresponding oxygen delignified pulps. Furthermore, as it mentioned earlier, the total COD levels are dependent on the initial kappa numbers. The COD compositions of eucalyptus and softwood effluents are significantly different, where the effluents from the eucalyptus pulps are more biodegradable. The compounds forming the kappa number in softwood and hardwood (especially eucalyptus) differ as well: in softwood, the kappa number mainly representative of lignin, whereas in eucalyptus, the hexenuronic acids (HexA) are a large contributor [22, 23]. In this regard, the most common way to remove the hexenuronic acids is in the early bleaching stages through hot acid hydrolysis (A) and hot chlorine dioxide bleaching (D_H) technologies [11, 22, 24].

6.2.2. *Chemical composition of the wastewater*

The two main types of bleaching methods in use are elemental chlorine free (ECF), when no molecular or gaseous chlorine is dosed in the bleaching, and totally chlorine free (TCF) bleaching [12]. ECF is dominating the bleached chemical pulp market. In 2012, ECF pulp production reached approximately 93% of bleached kraft pulp's world market share. TCF production has declined a little over the last 10 years [25].

Owing to the differences between both the bleaching technologies and chemical composition of the bleaching effluents, it is necessary to study in order to predict and understand the environmental impact associated, and consequently to develop the most suitable treatment that decreases effluent loads and toxicity. A significant number of studies pertaining to the chemical composition of bleaching effluents have been published. Several authors have worked in identifying the chemical compounds in filtrates. More than 500 organic compounds have been identified in bleaching effluents so far. Most compounds identified in bleaching effluents are derived from lignin or other wood components, such as extractives or carbohydrates [26].

The most important difference, when comparing softwood effluents with the eucalyptus effluents, is the higher lignin content in the former and the hexenuronic acid content in the latter [20].

Lignin degradation products were commonly considered as the major precursors of chlorinated compounds. However, the presence of monochlorinated compounds derived from glucuronxylans were identified to be the major components of chlorine dioxide bleaching filtrates of eucalyptus kraft pulps [27, 28].

Other important compounds found in the effluents are wood-derived components: resin acids, fatty acids, phytosterols and retene. Lipophilic hardwood extractives consist of a complex mixture of compounds such as sterols, long chain aliphatic acids and alcohols, waxes, glycerides and sterol esters. If high amounts of these compounds are found in kraft mill effluent, their origin is frequently the spills of black liquor and soap or black liquor transported with the pulp [14, 29].

6.2.3. *Molecular weight distributions*

Several authors [14, 30, 31] have worked in determining the molecular weight distribution of the components in the effluents. The importance of determining the molecular weight distribution comes from the fact that significant removal in the biological treatment system is achieved from the low molecular weight (LMW) material. Evidence of this is the increment in the proportion of organic compounds with high molecular weight after biological treatment. Improvements in the removal of high molecular weight material would lead to greater efficiency and improve the effluent quality. Traditionally, the separation between low molecular weight (LMW) and high molecular weight (HMW) is done at 1000 Da. Bleach kraft mill effluents have an extended molecular weight distribution; from diverse kinds of monomeric compounds to large and complex molecules with molecular weights between 10,000 and 30,000 g/mol. The molecular weight distribution depends on the raw material and the bleaching process used. For example, the average molecular weight of organic matter in hardwood kraft pulp effluents is lower than the corresponding softwood effluents [14].

The molecular weight fractions in the bleaching filtrates of oxygen delignified eucalyptus pulps were studied. The HMW fraction contributed to approximately 40% of the total effluent load of COD both in softwood and hardwood ECF bleached pulps production, and about 30–40% to TCF bleached pulps effluents [30, 31]. Additionally, the most remarkable differences between softwood- and hardwood-derived effluents are in the aromatic region. The aromatic lignin-derived structures such as syringyl and guaiacyl units are not important structural elements in HMW effluent materials from ECF bleaching of oxygen delignified hardwood kraft pulps, but are important in softwood HMW effluents [31, 32]. Similarly, the results show that all HMW effluents contained carbohydrates. The carbohydrates found in the examined HMW could have had oligosaccharides, polysaccharides or both present in the effluent, either in dissolved or colloidal form. As can be expected, the HMW hardwood kraft pulps fraction contained more carbohydrates (mainly xylan) than the corresponding samples from softwood kraft pulps. Concerning the presence of carboxylic acids, the HMW samples showed high levels of these groups. They were formed due to the oxidation of lignin structures in the bleaching process [30–32].

Regarding the low molecular weight (LMW) compounds, it can be broadly classified into three main classes: acids, phenolic compounds and neutral compounds. The phenolic compounds

http://dx.doi.org/10.5772/67537

and some of the acids are degradation products from lignin, while the resin acids, fatty acids, terpenes and sterols are residues of extractives presents in the raw material [14].

6.2.4. ECF and TCF wastewaters treatability

The biological treatment of the effluents from ECF and TCF is almost the same. There is a slight difference in the organic matter constitution among these bleaching effluents, but it is less than other parameters such as raw materials, effluents from the unbleached line, than the bleaching effluent itself.

TCF eucalyptus pulp produced an effluent with 3.5 times the BOD and twice the COD than ECF eucalyptus pulp effluent [30]. Similarly, TCF bleaching effluent had approximately twice the COD in softwood than the ECF effluents [33]. The larger amounts of COD and TOC in the TCF effluents can be explained because the bleaching reagents used in the TCF sequences (O_3, H_2O_2) are less selective towards residual lignin than the ClO_2 use in the ECF sequences. Bleaching of pulps with ozone is known to produce aldehyde and keto groups on carbohydrates, which are highly susceptible to oxidative degradation under alkaline conditions. An alkaline peroxide stage is used to further bleach ozone-treated pulps, resulting in an oxidative degradation of these carbohydrates and thus contributing to higher COD and TOC values in the TCF effluents. Moreover, the hardwood TCF effluents contained more carbohydrates (mainly xylan) than the ECF effluents. An explanation of these differences was that the process conditions in P-stage (long retention time under alkaline conditions) may favour dissolution of xylan from the pulp [30, 33].

However, while TCF effluent contains more dissolved organic matter, it is less coloured than ECF effluent, mainly because of the action of residual reagents (i.e. H_2O_2) in the TCF effluent. Normal values of colour at 525 nm in TCF effluents are 300 and 1300 C.U. in ECF effluents [31].

7. Kraft pulping: wastewater treatment

The typical pulp mill wastewater treatment should include primary treatment (neutralization, screening or sedimentation), principally to remove suspended solids, and biological/secondary treatment. The secondary treatment is mainly done to diminish the organic matter, which is removed by biological degradation, and is particularly useful for the removal of low molecular mass organic matter with a molecular weight of 800 Da or less. Some mills have tertiary treatment to further reduce toxicity, suspended solids, organics or colour [12, 13, 34].

Secondary biological treatment is applied in most types of pulp and paper mills. The most usual methods are activated sludge and aerated lagoons. Some variations of these systems include the use of filters and sequences reactors—Mobil Bed Bioreactor (MBBR) and Membrane Bioreactors (MBR). Sometimes anaerobic treatment is used followed by an aerobic biological stage [12, 18].

Aerated ponds and activated sludge methods are the most common treatment systems in pulp and paper industry. In an aerated pond, wastewater is treated through a combination of physical, biological and chemical processes. They have large residence times between 3 and 20 days, and consequently a large volume. They work with low microorganism concentration (low solids concentration) about 100–300 mg/L. These ponds use aeration devices to add oxygen to the wastewater (normally surface turbine aerators or bottom aerators) and mix the contents of the pond, thereby enhancing the microbial activity. However, due to low efficiency levels and the large surface required, the use of aerated lagoons has drastically diminished [12, 13, 34].

The largest secondary treatment system is activated sludge (60–75% of all the biological effluent treatment plants in pulp and paper industry use activated sludge systems); even in new plants. The advantages of the aerated activated sludge systems compared to the aerobic ponds are that they achieve high removal efficiencies, the process can be well controlled, requires less surface and the microorganisms are adapted to the receiving wastewater. The disadvantages are the high construction and operation costs (especially the energy cost of the aeration systems), the high rate of sludge production and the loss of efficiency due to bulking problems, and consequently, the need to add nutrients to avoid this problem. Sludge handling and nutrient dosage are additional to the energy cost, which is the major component contributing to the operational cost of the biological treatment of process effluents within the pulp and paper industry [34].

7.1. Characteristics of activated sludge treatment

Two main units of the activated sludge plant are the aeration basin and the sedimentation basin. In the aeration basin, the effluent is treated with a culture of microorganisms (the activated sludge), which is present in a high concentration. **Figure 7** shows a diagram of a pulp mill treatment with the activated sludge system. Activated sludge plants at kraft pulp mills have a retention time of about 15–48 h. The solids concentration in the activated sludge systems is typically 2000–6000 mg/L. The hydraulic residence time is 4–8 h for a conventional system and the cellular residence time (sludge age) is normally 5–15 days. Normal loads are between 0.05 and 0.1 kg BOD/kg sludge for extended aeration and 0.1–0.3 kg BOD/kg sludge for low load process. The common operating temperature is about 35–37°C and the dissolved oxygen (DO) concentration is 1.5–2.0 ppm. The nutrients concentration in relation to the organic matter is important in effluent treatment. Effluents from the wood processing industry generally have a BOD:N:P ratio of 100:(1–2):(0.15–0.3) and the addition of supplemental nutrients is normally required [13, 34].

The removal efficiencies reached vary according to the wastewater residence time and the operating conditions. Normal efficiencies figures are between 85 and 98% BOD_5 removal and 60–85% for COD removal. For AOX, the reduction is about 40–65%, 40–85% for phosphorus and 20–50% for nitrogen. The overall efficiency of TSS removal using primary and secondary treatment is about 85–90% [12].

7.2. Aerobic treatability of the different effluent fractions

The COD of treated effluent represents how effective a treatment technology is in its ability to remove the total organic material present in the influent. BOD measurements by themselves

http://dx.doi.org/10.5772/67537

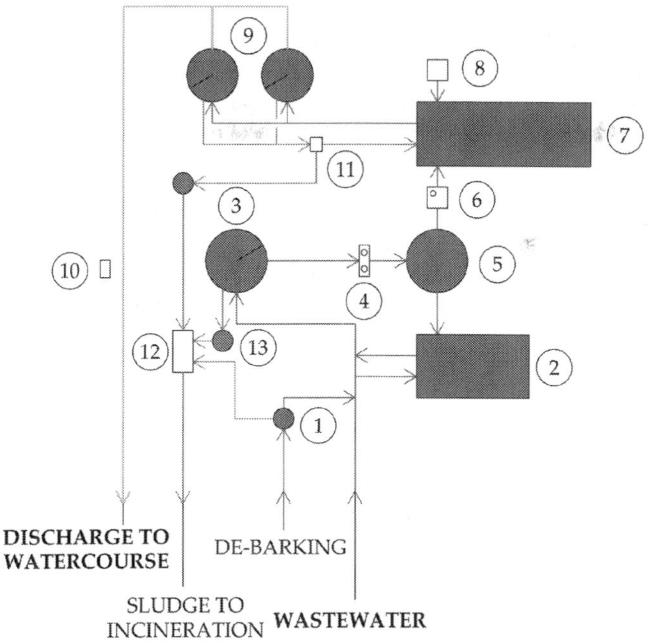

1-DE-BARK WATER CLARIFIER, 2-SPILL LAGOON, 3-CLARIFICATION, 4-COOLING TOWER, 5-EQUALISATION BASIN, 6-NEUTRALISATION AND NUTRIENT ADDITION, 7-AERATION BASIN, 8-COMPRESSOR STATION, 9-SECONDARY CLARIFICATION, 10-MEASUREMENT AND SAMPLING, 11-SCREEN PUMP, 12-SLUDGE HANDLING, 13-FIBRE SLUDGE THICKENER.

Figure 7. Diagram of a pulp mill treatment plant with activated sludge as biological treatment.

do not quantify the non-biodegradable or slowly biodegradable organic portion of the effluent. Moreover, studies seem to indicate that the residual colour in pulp mill effluents could be linked to the recalcitrant COD [35].

Recalcitrant organic matter is supposed to be partly responsible for long-term toxicity in receiving waters [21]. As discussed earlier, it is widely reported that the residual recalcitrant organic matter is composed predominantly by high molecular weight components, which are not metabolized due to its size. However, the contributions of high and low molecular weight fractions in bio-treated effluents are dissimilar [36]. In the LMW fraction, a large-scale removal of the chlorinated phenolic compounds, chlorinated resin acids and sterols occurs. In the HMW fraction, the carbohydrates are strongly affected; however, other compounds such as oxidized lignin were less affected [30].

Some findings are possible by comparing the high molecular weight (HMW) and low molecular weight (LMW) fractions of the acidic and alkaline filtrates post biological treatment [32]. In the alkaline filtrate, the COD and TOC in the HMW fraction increased after treatment. The same behaviour was observed with the AOX and lignin content in the acidic filtrate. This is attributable to the formation of soluble bacterial products or to the adsorption of the LMW into HMW matter [32, 35]. In the LMW filtrates, the COD/TOC decreased after biological treatment, as a result of the large removal of highly oxidized organic carbon. The colour increased in the HMW fractions of acid and alkaline filtrates. The biological treatment often leads to increased colour in ECF bleaching effluents due to the creation of new chromophores in the HMW fractions [13, 32].

7.3. Bulking problems in the activated sludge systems

Two critical operational aspects of an activated sludge plant are maintaining proper control of the dissolved oxygen (DO) concentration in the aeration tank and preserving a good settling sludge. Reduced settleablility results in poor plant performance, as it is difficult to maintain a low concentration of suspended solids in the plant effluent [13, 34]. Activated sludge plants that treat pulp and paper mill wastewaters seem to be particularly prone to this. There are several reasons for poor separation properties, such as filamentous bulking sludge, bulking due to excessive extracellular polymeric substances (EPS), production or formation of small flocs and dispersed biomass [37, 38]. In pulp mill wastewater, bulking is often due to the presence of filamentous bacteria. Common conditions that favour bulking are working at feeding loads ratios out of normal range, deficiencies in nitrogen and phosphorous species or in the level of DO [13]. In kinetic terms, the floc forming microorganisms have a competitive advantage at lower substrate concentrations because that allows the compounds to utilize oxygen and nutrients more efficiently than the not floc forming microorganisms [37].

The presence of filamentous bacteria was examined for two years in 15 French pulp, paper and board mills wastewater. The study of 25 bulking cases attributed the source in 10 cases to be COD hydraulic overloads, in 8 cases to deficient aeration and in 5 cases to nutrient deficiency [39].

8. Partial closure in water circuits

The current market and environmental demands facing pulp and paper mills are the increased closure of the plant circuits and a further reduction or elimination of the wastes produced. The concept of a closed loop mill aims to eliminate discharges to the aquatic environment, recycle and reuse all possible solid and liquid process wastes, and reduce air emissions to the lowest possible quantity and toxicity. However, until today, no kraft mills are operating with complete closure and complete reutilization of the effluents. The most important problem experienced in mills that try to operate for long periods with zero discharge was corrosion caused by chlorides in a number of positions. Nevertheless, great progress has been made in minimizing impacts associated with pulp mill effluents. Water circulation closure methods include dry debarking, effective liquor spilling control, closed screening and washing, condensate stripping and other methods to minimize the loss of wood-derived organic mat-

http://dx.doi.org/10.5772/67537

ter. Extended and oxygen delignification can significantly reduce bleach plant effluent loads from kraft pulp mills. The bleach plant is the most important source of effluent within a pulp mill and the chlorinated effluents are more complicated to reutilize within the mill. For this reason, an important trend in bleaching development is to reduce volumes and decrease the effluent loads, especially of chlorinated compounds [40–42].

Up to now, a complete water closed circulation is not available; nevertheless, a partial closure of the water circuits is possible. This can be done segregating the acid and alkaline effluent streams and recirculating the liquids countercurrently from the last bleach stage through the sequence to the brown stock washer. The alkaline effluent could be used for washing the pulp in the unbleached part of the process.

9. Conclusions

The pulp and paper industry has been considered a large consumer of wood, energy and water, and an important contributor of pollutant discharges to the environment (air, water courses and soil). However, the last decades have seen a lot of effort in creating solutions such as generating less pollutant wastewaters and reducing the amount and load of the emissions to the environment. The implementation of several measures like the dry debarking of wood, the introduction of extended cooking and oxygen delignification, the reuse of condensates, improvements in washing efficiency and especially the total substitution of chlorine, has brought a significant reduction in effluent flows and in the chlorinated and organic loads generated within the mill. In addition, the introduction of end-of-pipe secondary, and even tertiary, treatments have reduced large amounts of pollutant loads to the environment. However, the need for tertiary treatment is not yet well proven; while it purifies the effluent, the energy costs are high and even forms sludge.

Effluent characteristics are dependent on the production process and the raw materials. ECF eucalyptus pulp production is increasing appreciably but not much information on its effluents is available. The main difference between softwood and eucalyptus pulps is in the kappa number: the kappa number is mainly formed by lignin content in softwood pulp, and the Hexenuronic acids are important contributors to kappa number in eucalyptus pulp. Hence, the bleaching conditions for eucalyptus are less severe and consequently the effluents characteristics are different. Eucalyptus bleaching effluents have lower COD, AOX and colour content and higher biodegradability than the softwood effluents.

The environmental impact of effluent loads and the appropriate treatment can be determined by studying the chemical composition and molecular weight distribution of the bleaching effluents. The HMW in hardwood bleaching wastewaters constituted an important but not prevailing fraction of the wastewater composition (30–65% of the total). The hardwood HMW fraction is mainly composed of non-aromatic structural compounds.

Aerated activated sludge is the most common treatment system in pulp mills. BOD_5 removals of 85–98% and COD removals of 60–85% are normally achieved with these systems. For AOX, the reduction is about 40–65%, 40–85% for phosphorus and 20–50% for nitrogen. Bulking problems are common in these systems mainly due to nitrogen deficiencies and phosphorous concentration or the level of DO.

Nowadays, plants that apply the best available technologies have their emissions controlled and present minimum environmental impact at the receiving waters.

The new developments are in the way to close even more the internal circuits in the plant, to reduce the flow discharged. Membrane technologies and similar technologies may be key in this regard.

Author details

María Noel Cabrera

Address all correspondence to: ncabrera@fing.edu.uy

Universidad de la República, Engineering School, Chemical Engineering Institute, Forest Process Engineering, Montevideo, Uruguay

References

[1] Diesen M. Economics of the pulp and paper industry. 1st Ed. Helsinki (Finland): Paperi ja Puu Oy; 2007, 186 p.

[2] FAO. Forest products yearbook, 1979–1990 [Internet]. Rome (Italy); 1992, p. 392. Available from: http://www.fao.org/forestry/statistics/80570/en/ [Accessed 2009 Oct 8]

[3] FAO. Forest products yearbook, 1991–1995 [Internet]. Rome (Italy); 1996. Available from: http://www.fao.org/forestry/statistics/80570/en/ [Accessed 2009 Oct 15]

[4] FAO. Forest products yearbook, 1996–2001 [Internet]. Rome (Italy); 2003. Available from: http://www.fao.org/forestry/statistics/80570/en/ [Accessed 2009 Sep 8]

[5] FAO. Forest products yearbook, 2002–2006 [Internet]. Rome (Italy); 2008. Available from: http://www.fao.org/forestry/statistics/80570/en/ [Accessed 2010 Oct 15]

[6] FAO. Forest products yearbook, 2009–2013 [Internet]. Rome (Italy); 2015. Available from: http://www.fao.org/forestry/statistics/80570/en/ [Accessed 2016 Jul 11]

[7] Smook GA. Handbook for pulp & paper technologists. 3rd ed. Tappi Press; Vancouver--B.C.; 2003, 425 p.

[8] Sixta H, editor. Handbook of pulp. 1st Ed. Weinheim: WILEY-VCH Verlag GmbH &Co. KGaA; 2006, 1369 p. doi:10.1002/9783527619887.

[9] Stenius P. Forest products chemistry. 1st Ed. Gullichsen J, Paulapuro H, editors. Fapet Oy; Helsinki; 2000, 348 p. i

[10] Gullichsen J. Fiber line operations. In: Gullichsen J, Fogelholm CJ, editors. Chemical pulping. Fapet Oy; Helsinki; 1999, p. 693.

[11] Colodette JL, Gomes CM, Rabelo M. Progess in eucalyptus kraft pulp bleaching. In: 2nd International Colloquium on Eucalyptus Pulp (2ICEP); Concepcion-Chile; 2005, pp. 1–18. Available from: http://www.eucalyptus.com.br/icep02/jorge_colodette.pdf. [Accessed 2009 Oct 15]

[12] European Commission, Joint Research Centre, Institute for Prospective Technological Studies. Integrated Pollution and Prevention Control. Best Available Techniques (BAT) reference document for the production of pulp, paper and board. 2015, Available from: http://eippcb.jrc.ec.europa.eu/reference/BREF/PP_revised_BREF_2015.pdf. [Accessed 2016.01.20]

[13] Springer A. Industrial environmental control: Pulp and paper industry. 3rd Ed. Atlanta, GA (USA): TAPPI Press; 2000.

[14] Lehtinen K. Relationship of the technical development of pulping and bleaching to effluent quality and aquatic toxicity. In: Borton DL, Hall T, Fisher R, Thomas J, editors. Pulp & paper mill effluent environmental fate & effects. Destech Publications Inc.; Lancaster-PA; 2004.

[15] Munkittrick KR, Servos MR, Carey JH, Van Der Kraak GJ. Environmental impacts of pulp and paper wastewater: Evidence for a reduction in environmental effects at North American pulp mills since 1992. Water Sci Technol. 1997;**35**(2–3):329–38.

[16] Thompson G, Swain J, Kay M, Forster CF. The treatment of pulp and paper mill effluent: A review. Bioresour Technol. 2001;**77**(3):275–86.

[17] Rodgers JH, Thomas JF. Evaluations of the fate and effects of pulp and paper mill effluents from a watershed multistressor perspective: Progree to date and future opportunities. In: Borton DL, Hall T, Fisher R, Thomas J, editors. Pulp & paper mill effluent environmental fate & effects. DEStech Publications Inc.;. Lancaster- PA; 2004

[18] International Finance Corporation (IFC). Environmental, health, and safety guidelines pulp and paper mills. 2007. Available from: http://www.ifc.org/wps/wcm/connect/6f13e78048855398afb4ff6a6515bb18/Final%2B-%2BPulp%2Band%2BPaper%2BMills.pdf?MOD=AJPERES [Accessed 2010 April 5]

[19] Dahl O. Process modifications to reduce effluent loads. In: Dahl O, editor. Environmental management and control. 1st Ed. Paperi ja Puu Oy; Helsinki; 2008, pp. 70–84. ISBN: 978-952- 5216-30-1.

[20] Herstad Svärd S, Basta J, Wäne G, Jour P. Comparative characterization of the bleach plant effluents from modern ECF bleaching—eucalyptus and softwood pulps. In: 30th pulp & paper annual meeting, ABTCP. 3–7 November 1997, Sâo Paulo, Brazil; pp. 49–62.

[21] Yousefian S, Reeve D. Classes of compounds responsible for COD and colour in bleached kraft mill effluents. In: International environemental conference & exhibit: Setting the environmental course for the 21st century. 6–10 May 2000, Denver, CO (USA); pp. 657–65.

[22] Teleman A, Hausalo T, Tenkanen M, Vuorinen T. Identification of the acidic degradation products of hexenuronic acid and characterisation of hexenuronic acid-substituted xylooligosaccharides by NMR spectroscopy. Carbohydr Res. 1996;**280**(2):197–208. doi:10.1016/0008-6215(95)00309-6.

[23] Ventorim G, Oliveira KD, Colodette JL, Da Costa MM. Effect of pulp kappa number, lignin and hexenuronic acid contents on oxygen delignification performance. Sci For Sci. 2006;**71**:87–97.

[24] Ragnar M, Lindström ME. A comparison of emerging technologies. Pap ja Puu. 2004; **86**(1):39–44.

[25] Alliance for Environmental Technology. Trends in world bleached chemical pulp production: 1990–2012. AET Reports. 2013. Available from: http://www.aet.org/science_of_ecf/eco_risk/2013_pulp.html [Accessed 2016 Aug 10].

[26] Mc Kague B, Carlberg G. Effluent characteristics and composition. In: Dence C, Reeve D, editors. Pulp bleaching-principles and practices. TAPPI Press; Atlanta- GA; 1996, pp. 751–65. ISBN:0-89852-063-0.

[27] Freire CSR, Silvestre AJD, Pascoal Neto C. Carbohydrate-derived chlorinated compounds in ECF bleaching of hardwood pulps: Formation, degradation, and contribution to AOX in a bleached kraft pulp mill. Environ Sci Technol. 2003;**37**(4):811–4. doi:10.1021/es0200847.

[28] Freire CSR, Silvestre AJD, Neto CP, Silva AMS, Evtuguin DV, Cavaleiro JAS. Easily degradable chlorinated compounds derived from glucuronoxylan in filtrates from chlorine dioxide bleaching of eucalyptus globulus kraft pulp. Holzforschung. 2003;**57**(1):81–7. doi:10.1515/HF.2003.013.

[29] Freire CSR, Silvestre AJD, Pascoal Neto C. Oxidized derivatives of lipophilic extractives formed during hardwood kraft pulp bleaching. Holzforschung. 2003;**57**(5):503–12. doi:10.1515/HF.2003.075.

[30] Dahlman O, Reimann A, Stromberg L, Mörck R. High molecular weight effluent materials from modern ECF and TCF bleaching. Tappi J. 1995;**78**(12):99–109. ISBN: 978-952-5216-30-1.

[31] Mounteer AH, Passos FML, Borges AC, Silva DO. Detecting structural and functional differences in activated sludge bacterial communities originating from laboratory treatment of elementally and totally chlorine-free bleaching effluents. Can J Microbiol. 2002;**48**(3):245–55.

[32] Souza L, Mounteer AH, Silva C., Dalvi L. A study on biological removal of recalcitrant organic matter in eucalyptus kraft pulp ECF bleaching filtrates. In: Tappi International Environmental Conference, CD-ROM. Atlanta, GA (USA): Tappi Press; 3–5 May 2003, pp. 1–8.

[33] Cates D, Eggert C, Yang J, Eriksson K. Comparison of effluents from TCF and ECF bleaching of kraft pulps. Tappi J. 1995;**78**(12):93–8.

[34] Hynninen P. Effluent treatment. In: Dahl O, editor. Environmental management and control. 1st Ed. Paperi ja Puu Oy; Helsinki; 2008. pp. 86–116.

[35] Konduru RR, Liss SN, Allen DG. Recalcitrant organics emerging from biological treatment of kraft mill effluents. Water Qual Res J Canada. 2001;**36**(4):737–57. doi:10.1139/w02-006.

[36] Jokela JK, Laine M, Ek M, Salkinoja-Salonen M. Effect of biological treatment on halogenated organics in bleached kraft pulp mill effluents studied by molecular weight distribution analysis. Environ Sci Technol. 1993;**27**(3):547–57. Available from: http://pubs.acs.org/doi/abs/10.1021/es00040a014

[37] Andreasen K, Agertved J, Petersen JO, Skaarup H. Improvement of sludge settleability in activated sludge plants treating effluent from pulp and paper industries. Water Sci Technol. 1999;**40**(11–12):215–21.

[38] Tsang Y, Chua H, Sin S, Tam C. A novel technology for bulking control in biological wastewater treatment plant for pulp and paper making industry. Biochem Eng J. 2006;**32**:127–34. doi:10.1016/j.bej.2006.08.014.

[39] Fourest E, Craperi D, Deschamps-Roupert C, Pisicchio JL, Lenon G. Occurence and control of filamentous bulking in aerated wastewater treatment plants of the French paper industry. Water Sci Technol. 2004;**50**(3):29–37.

[40] Axegård P, Carey J, Folke J, Gleadow P, Gullichsen J, Pryke D, et al. Minimum impact mills : Issues and challenges. In: Tappi minimum effluent mills symposium. San Francisco, CA (USA): TAPPI Press; 1997. Available from: http://infohouse.p2ric.org/ref/18/17842.pdf. [Accessed 2008.10.15]

[41] Fiskari J, Toikkanen L, Gullichsen J. Experiments on bleaching effluent minimization. In: TAPPI pulping/process and product quality conference. Boston, MA (USA); 5–8 November 2000, pp. 771–8.

[42] Stratton SC, Gleadow PL, Johnson AP. Pulp mill process closure: A review of global technology developments and mill experiences in the 1990s. Water Sci Technol. 2004;**50**(3):183–94.

Chapter 8

Molecular Biomonitoring of Microbial Communities in Tannery Wastewater Treatment Plant for the Removal of Retanning Chemicals

Adey Feleke Desta, Joyce Nzioki,
Solomon Maina and Francesca Stomeo

Additional information is available at the end of the chapter

http://dx.doi.org/10.5772/67349

Abstract

This chapter focuses on culture-independent characterization and monitoring of microbial communities in tannery wastewater treatment system, with special reference to the degradation of two xenobiotic chemicals used in retanning processes. Molecular survey of a tannery wastewater treatment system through metagenomic and metatranscriptomic approaches revealed a diverse microbial community in each component of the treatment system with high gene copies for enzymes involved in the degradation of cyclic aromatic compounds such as nitrotoluene. A combination of flow cytometry and molecular fingerprinting methods was used in a lab-scale reactor to monitor the dynamics of the microbes in the sludge and the fate of two retanning chemicals. The identified key microbial communities for the removal of the two xenobiotic chemicals belong to members of the group *Proteobacteria* and the phylum *Bacteroidetes*.

Keywords: tannery, retanning chemicals, bacteria, biomonitoring, wastewater, flow cytometry, fingerprinting

1. Introduction

1.1. The leather industry

The leather industry in developing nations is a sector in continuous growth but leaving behind the toxic pollutants in the environment. The economy of Eastern African countries is predominantly agricultural where the livestock subsector plays a substantial role. Livestock is an integral part of the national agricultural wealth of Eastern African countries serving as sources of power, meat, milk, egg, hides and skins, manure, and other products. Hides and skins, though by-products of animals, have been contributing greatly to the export earnings from the livestock sector since ancient times.

In most parts of sub-Saharan Africa and Asia, tanning is a family business, carried out in small- to medium-scale semi-mechanized units. Tanneries owned by different individuals are frequently grouped tightly in clusters which used to be nonresidential areas. Most of the tanning facilities are strategically located near to rivers and small streams so as to discharge their large amount of heavily polluted wastewater directly to these water bodies. Considering a case study in Ethiopia, the Awash River is used as inputs for small- and large-scale farms of fruits, vegetables, and sugarcane, yet experiencing a significant water quality deterioration. The discharge of properly untreated tannery effluent has caused severe pollution affecting surface and underground water resources, farms irrigated by such water, people working in the farms, and consumers of the farm products, not to mention the aquatic ecosystem.

1.2. The leather manufacturing process

The production of leather involves the whole process of converting raw hides or skins into useful commodities such as shoes and garments from the meat industry [1]. Hides and skins are processed to react with various chemical substances that prevent them from putrefaction to make them resistant to wetting and keep them supple and durable [2, 3]. It has been reported by Khan and colleagues [4] that about 130 chemicals are used in the entire process of leather production. The production process is generally divided into four main categories, namely, the beamhouse, tanning, retanning, and finishing processes. In this chapter, we will focus on pollutants of the retanning process and their fate during biological treatment processes.

1.3. The retanning process: a closer look

Retanning, also called post-tanning operation, involves neutralization and fat liquoring to improve the feel and handle of leather and provide frame retarding, water, and abrasion resistance properties [5, 6]. Retanning is carried out by employing various substances such as phenolic and naphthalene resins, melaminic resins, acrylic resins, and polymers [6].

1.3.1. Melaminic resins as retanning agents

Melaminic resins are condensation products from formaldehyde with amino and amido compounds, such as urea, melamine, and cyanamide (dicyandiamide). The amine resins are polymers synthesized by condensation of urea, formaldehyde, and melamine (2,4,6-tri-amino-1,3,5-triazine) [7]. The formaldehyde undergoes an addition reaction with amino group of urea or melamine with the formation of N-methylol groups. Urea-formaldehyde resins are synthesized and chemically modified by reaction with a sulfating agent to form a sulfonated soluble product. Regarding melamine, the methylol groups can react with amino or other methylol groups to form methylene or ether bridges based on the reaction scheme for melamine as depicted in **Figure 1**. These resins give light colored leathers with good resistance.

Due to their availability, melaminic resins are among the widely used chemicals in industries processing leather to the retanning and finishing steps [7]. The trade name Retanal MD-80 refers to melamine-formaldehyde resin used in retanning of hides and skins.

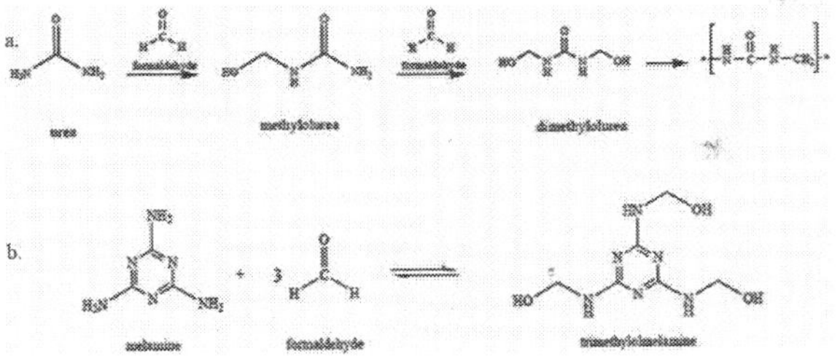

Figure 1. Condensation of urea (a) and melamine (b) using formaldehyde (after Ref. [7]).

1.3.2. Phenolic and naphthalene resins as retanning agents

Phenolic and naphthalene resins are polymers synthesized using phenolic, naphthalene, and their derivatives condensed with urea and formaldehyde. The synthesis reaction which is patented by BASF in 1913 involves reaction of the basic phenolic and/or naphthalene constituents under acidic conditions which results in attachment of the aromatic compounds to one another with the aid of formaldehyde through methylene bridges. Then, they are adjusted to the optimum degree of condensation by making them binuclear or trinuclear and made water soluble by sulfonation or sulfomethylation which are finally adapted by buffering to meet the application requirements (**Figure 2**) [6, 8]. When used on chrome-tanned leather, they specifically impart it to a soft fullness and relaxed grain. These characteristics of mellowness and softness are very desirable in gloves, garment, and soft-type leathers [9].

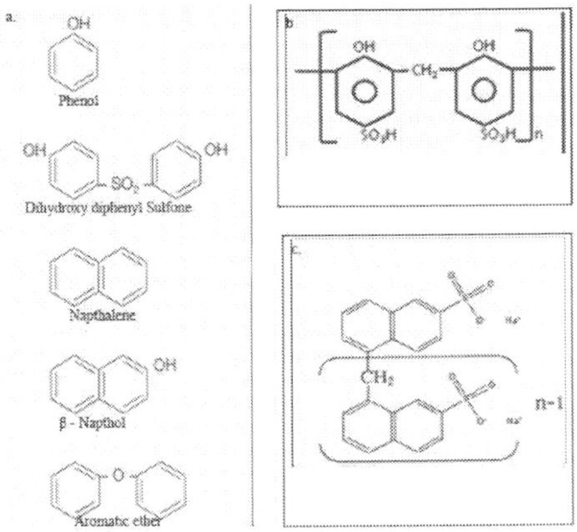

Figure 2. (a) Basic constituents of phenolic and naphthalene resins, (b) structure of phenol formaldehyde condensate, and (c) structure of naphthalene formaldehyde condensate (after Refs. [6, 10]).

COD (mg/l)	BOD (mg/l)	NH_4 (mg/l)	Cr (mg/l)	S^{2-} (mg/l)	TS (mg/l)	SS (mg/l)	VSS (mg/l)	pH	TDS (mg/l)	Reference
2250±565	1000±88	–	0.027±0.075	–	–	92±36	–	6.14±1.1	–	[14]
3700	1470	180	–	440	–	2690	1260	7.4	–	[15]
4800±350	–	225±18	95±55	–	10,266±1460	2820±140	1505±90	7.06 ± 0.26	18,800–19700	[16]
1320–54,000	840–18620	–	41–133	800–6480	–	220–1610	–	–	–	[17]
4100–6700	680–976	144–170	41,623	–	–	600–955	–	7.0–8.7	–	[18]
8000	930	–	11	228	–	2004	1660	–	15,152	[19]
2200	–	–	–	–	–	5300	1300	7.7	36,800	[20]
11123±563	2983±259	122±8	32±6	630±67	–	–	–	10.8-0.1	6646±557	[21]
2155	–	166	50.9	35.6	–	915	578	7.79	–	[22]
3114	1126	33	83	55	18,884	1147	–	10.5	17,737	[23]
2426	–	335	29.3	286	–	–	–	7.7	–	[24]
5650	–	–	–	–	19,755	5025	–	8.2–8.5	14,750	[25]

The first column lists the parameters used in the different studies. Parameters with "–" means data are not available.

Table 1. Characteristics of tannery wastewater based on studies from different countries and treatment systems.

1.4. Characteristics of tannery wastewater

The tanning process consumes high amount of water, estimated to be 34–56 m^3 of water per ton of hide or skin processed [11]. Out of the total water consumed, 85% is discharged as a wastewater [12]. Interestingly, only 20% of the wet-salted hides/skins are converted into commercial leather, 25% becomes chromium-containing leather waste, and the remainder becomes non-tanned waste or is lost in wastewater as fat, soluble protein, and solid suspended pollutants [13]. Therefore, environmental pollution remains to be a serious problem in the leather sector.

The characteristics of the wastewater vary considerably from tannery to tannery depending on the size of the tannery, the chemicals used for the specific process, the amount of water used, and the type of final product produced by a tannery. The variations of effluent characteristics also occur through each working day in a tannery. According to Calheiros et al. [14], average COD and pH analyzed in 1 day were 2010 mg/l (±516) and 6.98 (±0.05), respectively, whereas 2068 mg/l (±446) and 7.93 (±0.08), respectively, in another day. **Table 1** summarizes the pollution load discharged from individual tannery processing operations.

Most of the studies on pollution load of tanneries do not include chemicals that are involved in the process after the tanning step. This is partly because the pollution load of the chemicals used in the retanning process is included in some parameters such as COD and TDS. The other reason is the absence of fast and cheap method to detect these specific chemicals. Reemtsma et al. [26] reported the presence of benzothiazoles in tannery wastewater in three forms, benzothiazole (BT), methylthiobenzothiazole (MTBT), and monobenzothiazole (MBT), with a dominance of MBT at a concentration of 3.3–6.9 μmol/L. These compounds have been detected in tannery wastewater samples by Fiehn et al. [27] in concentrations of 655 μg/L MBT, 10.5μg/L BT, and 39 μg/L of MTBT. A report by UNIDO [28] indicated that only 22% of all the chemicals used for post-tanning and finishing process is taken up and remained in/on the leather, whereas from the remaining waste chemicals (88%), 23% belongs to fat liquors and 20% to dyestuffs.

In this study, we explore the microbial community in the different components of a treatment plant and expressed genes for the target chemicals Basyntan and Retanal. In addition, we decipher the key microbial subcommunities responsible for the degradation of our target post-tanning chemicals.

2. Materials and methods

2.1. Reactor setup and sampling

The data shown in this chapter are from a study conducted on a pilot-scale biological wastewater treatment plant installed in the premises of a privately owned tannery in Modjo town, Ethiopia, 70 km south of the capital Addis Ababa. The system consists of two anaerobic reactors each with volume of 25 m^3: an aerobic reactor with a volume of 50 m^3 and subsurface-flow constructed wetland vegetated with the perennial grass *Phragmites australis* (Cav.) (**Figure 3**). Performance of the treatment system was evaluated by taking samples of the influent and the treated effluent water and analyzing the different physicochemical parameters following the procedure in APHA [29].

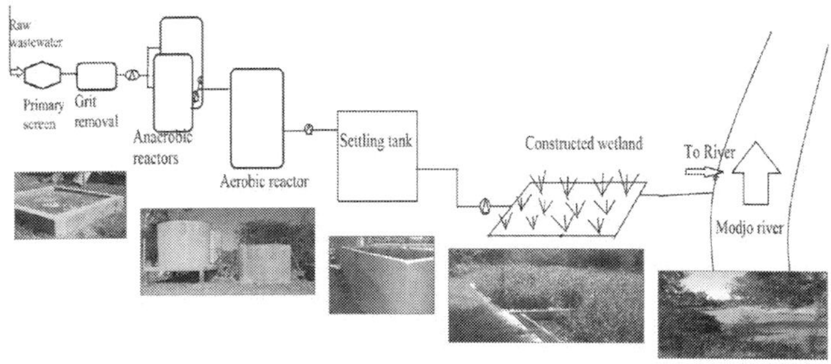

Figure 3. Schematic presentation of the pilot tannery effluent treatment site comprising anaerobic-aerobic reactors integrated with constructed wetland system.

2.2. Metagenomic and metatranscriptomic analyses

Sludge and sediment samples were taken from the anaerobic, aerobic, and different parts of the constructed wetland. The extraction of DNA and RNA was carried out using Zymo ZR® kit for DNA and Zymo ZR® kit for RNA (Zymo Research, CA, USA), respectively. Shotgun sequencing of the metagenome was conducted by means of Illumina Nextera XT® protocol. Total RNA was sequenced following the Illumina TruSeq® RNA preparation protocol.

The quality of the generated DNA and RNA reads was checked using FastQC toolkit [30]. FASTX-Toolkit was used to dereplicate, screen for ambiguous reads, and trim based on the cutoff value of Phred score >20 [31]. Assembly of the trimmed DNA and RNA reads was performed using Velvet (v 1.1) [32] and Trinity (v 2014) [33], respectively. Ribosomal RNA was removed using the riboPicker software (v 0.4.1) [34]. Binning and normalization were performed using an in-house Perl script. Taxonomic identification was done using BLASTN for the metagenome contigs and BLASTX for the metatranscriptome against a local download of NCBI nonredundant GenBank database. A set of contigs from the metatranscriptomic dataset were analyzed for the frequency of various identified genes, and Blast2GO (v 1) [35] was employed for the annotation of the genes.

2.3. Monitoring of microbial communities for the degradation of retanning chemicals

A bench-scale sequencing batch reactor (SBR) mimicking the treatment system was set up to analyze the dynamics of microbial community and its functional significance for the removal of the various pollutants in the wastewater. The SBR was operated continuously in cycles of around 72 hours with the fill, react, settle, and draw cycles as depicted in **Figure 4**. A number of abiotic parameters including liquid chromatography-based analysis of the two retanning chemicals (Basyntan and Retanal) were measured at each batch throughout the entire running period. Similarly, flow cytometry-based quantification and sorting of sludge microbial community stained with DAPI were carried out using the MoFlo cell sorter (DakoCytomation, Fort Collins, CO). The sorted cells were processed for taxonomic identification of the different subcommunities using T-RFLP and clone library-based 16S rRNA sequence analysis described in Koch et al. [36]. Correlation analyses between the abiotic parameters and the gated subcommunities were done by Spearman's rank-order correlation coefficient using the program R (http://CRAN.R-project.org) Version 2.14.0.

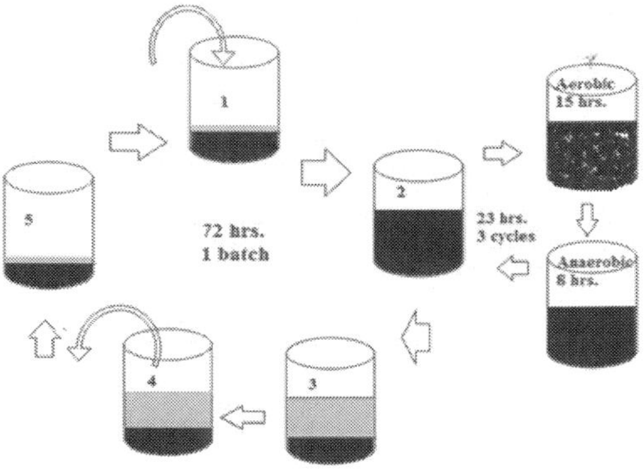

Figure 4. Schematic presentation of the different phases of the lab-scale sequencing batch reactor.

3. Results and discussion

3.1. Performance of the treatment system

Based on the physicochemical analysis, the untreated wastewater was characterized by its high concentration of sulfate, ammonia nitrogen, total suspended and dissolved solids, as well as high biological and chemical oxygen demands (BOD and COD). The high pH also indicated the alkalinity of the wastewater. Performance of the treatment system with regard to the removal of priority pollutants ranged between 70 and 99% (**Table 2**). The effluent parameters obtained for the COD, sulfate (SO_4^{2-}), sulfide (S^{2-}), nitrate (NO_3), and ammonia nitrogen (NH_3-N) were in line with the provisional emission limit values set for tannery effluents in Ethiopia which are 500 mg/l for COD, 1 g/l for SO_4^{2-}, 1 mg/l for S^{2-}, 20 mg/l for NO_3, and 30 mg/l for NH_3.

3.2. Metagenomic and metatranscriptomic analyses of tannery treatment system

Shotgun metagenomic analysis of the pilot reactors revealed the presence of seven phyla in the anaerobic reactor and eight phyla in the aerobic and the constructed wetland areas. The most abundant bacterial phyla in the anaerobic and aerobic reactors belonged to phylum *Firmicutes* and *Proteobacteria*, respectively. In the wetland, members of the phyla *Proteobacteria*, *Chlorobi*, and *Chloroflexi* were dominant (**Figure 5**).

A closer look into the dominant phylum *Firmicutes* showed that the genera *Bacillus*, *Clostridium*, and *Tissierella* were relatively the most abundant genera in the anaerobic system; these microorganisms have been implicated in the degradation of aromatic hydrocarbons in other tannery wastewater treatment systems [37].

Parameter	Influent	Effluent	% Removal
TN	245.25 ± 76	62.75 ± 14	74
SO_4	800 ± 505	35 ± 61	96
TP	15.33 ± 1	4.23 ± 2	72
S^{2-}	55.50 ± 6	4.91 ± 3	91
NO_3	310 ± 203	40.25 ± 28	87
NO_2	2.08 ± 3	0.03	99
NH_3	287.70 ± 178	44.28 ± 26	85
COD	12,547.50 ± 3910	395 ± 139	97
BOD	4886.26 ± 266	308.91 ± 24	94
TDS	9470.50 ± 1335	2593.69 ± 344	73
TSS	1155 ± 203	92 ± 11	92
VSS	27,482.75 ± 197	2272.75 ± 724	92
Total Cr	27.25 ± 3	0.95	97
pH	10.40 ± 0.3	7.66 ± 0.1	

Key: TN, total nitrogen; TP, total phosphorous; TDS, total dissolved solids; TSS, total suspended solids; VSS, volatile suspended solids; total Cr, total chromium (Source: Desta et al. [37])

Table 2. Average characteristics of the influent and effluent wastewaters of the integrated treatment system at the time of sludge sampling (concentrations are in mg/l, except for pH).

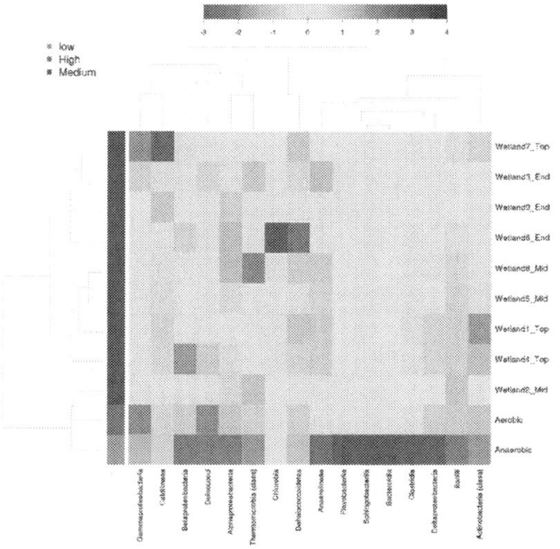

Figure 5. Relative abundance of bacteria as classified in phylum level. Sample sites were classified based on the concentration of salt (measured as TDS) and qualitatively designated as high, medium, and low levels.

The phyla *Bacteroidetes, Cyanobacteria, Actinobacteria,* and *Chloroflexi* follow the next levels of abundance in the anaerobic reactor, with members identified in the degradation on both priority nutrients and synthetic aromatic compounds [36, 37].

In the aerobic reactor, members of the phyla *Cyanobacteria* and *Deinococcus-Thermus* were the most abundant bacterial groups. The genus *Deinococcus* was more abundant in the aerobic reactor than in any other part of the treatment system. Members of the class *Betaproteobacteria* such as the genera *Burkholderia, Rhodocyclus,* and *Nitrosomonas* were identified from the aerobic system and are inferred to be involved in ammonia oxidation and aromatic compound degradation [36, 37].

Metatranscriptomic analysis of biological samples from the anaerobic reactor of the treatment system revealed the presence of genes coding enzymes involved in the degradation of nitrotoluene, chlorocyclohexane, toluene, and benzoate, apart from the enzymes for common anabolic and catabolic pathways (**Figure 6**). Relatively higher number of expressed genes were detected for nitrotoluene degradation coded for the enzymes DNT dehydrogenase (EC 1.2.99.2) and DHAT reductase (EC 1.8.99.3). These enzymes are implicated in the degradation of compounds such as nitrotoluene and related aromatic compounds.

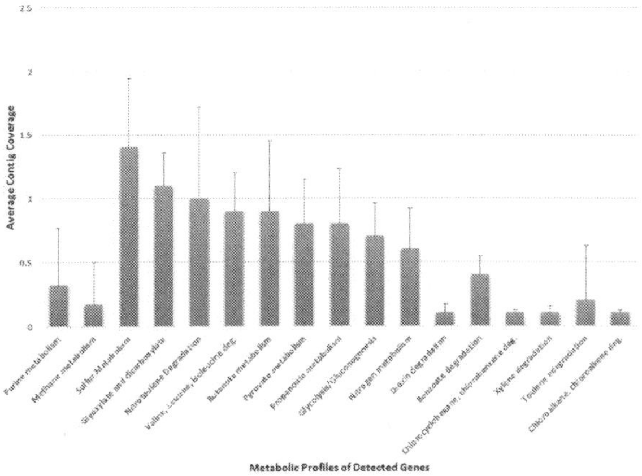

Figure 6. Average contig coverage for 17 common metabolic gene anaerobic reactors of tannery WWTP. X-axis, The genes involved in the metabolic pathways, and Y-axis, the average contig coverage. Error bars represent the standard error.

3.3. Dynamics and functional characterization of microbial communities

Flow cytometric characterization of bacterial community in the sludge of the reactor followed up at bench-scale sequencing batch reactor (SBR) revealed the dynamics, succession, and shift of the microbial subcommunities during the course of reaction, with typical patterns in each batch. Starting from the first batch of the operation of the SBR, changes expressed as shift of clusters in the x- and y-axes were observed in each batch of reaction of the SBR, indicating increase in cell size and proliferation activity of the microbial communities over the whole running period of the SBR. Based on visual evaluation of the histograms of the dot plots, a gate template was created representing 30 clusters during the 14 batches of the reactor run (**Figure 7**). From each gate, cell abundance over the entire reaction period was evaluated.

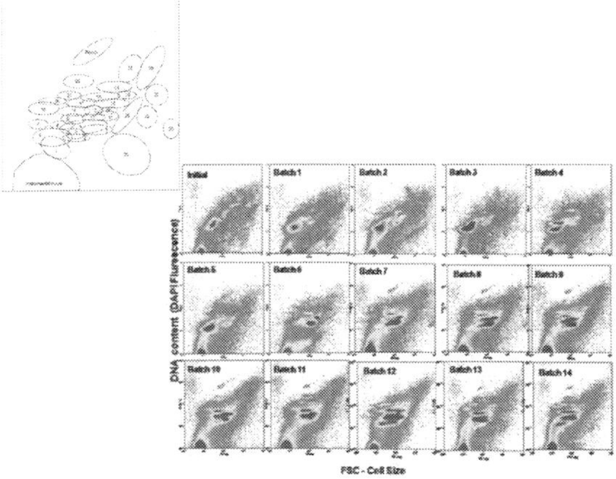

Figure 7. Bacterial community dynamics of tannery wastewater running in sequencing batch reactor (SBR) for 45 days. The first dot plot (initial) refers to the bacterial community in the acclimatized sample used to seed the reactor at the beginning of the SBR. The gate template (top-left box) which is used as the basis for fingerprinting of the different cell types.

Correlation analysis of bacterial cell abundance in each gate with the 13 measured abiotic parameters revealed positive correlations ($p<0.05$) between removal of the retanning agents and bacterial groups in gates G6, G12, and G20. Considering the different UPLC-based peaks of Basyntan, highly positive correlation was found specifically between peak 1 (ΔB1) and peak 3 (ΔB3) of Basyntan and the cells in G21 and G23. The correlations between the rest of the retanning agents (ΔB2 and ΔR) with the cells in G21 and G23 were still found to be positive ($p< 0.05$), suggesting the possible role of the cells in G21 and G23 for the biodegradation of Retanal and all the components of Basyntan. In order to have a closer look at the clusters and identify the consistent members in the flow cytometric pattern from each batch of the SBR, eight of the 30 gated subcommunities, namely, G1, G2, G6, G12, G14, G16, G20, and G21 were sorted to analyze the composition and abundance of bacteria in each sorted gate. From all the sorted gates, eight bacterial families and classes belonging to the phyla *Proteobacteria*, *Bacteroidales*, and *Bacteroidetes* were identified using terminal restriction fragments (T-RFs). Out of the eight gated clusters, gate 14 (G14) contained the smallest portion of the sorted bacterial community with predominant members belonging to *Proteobacteria* (6%) and showed strong positive correlation ($p< 0.01$) with the degradation of Basyntan and Retanal (**Figure 8**).

The gates 16, 20, and 21 (G16, G20, and G21) which showed positive correlations with retanning chemicals degradation were dominated by members of the phylum *Bacteroidetes* constituting 13,

23, and 66%, respectively. *Rhodocyclaceae* (11%), *Brucellaceae* (10%), and unclassified *Proteobacteria* (8%) were the second, third, and fourth abundant groups identified in gate 20 (G20), respectively. The most abundant cells belonged to *Rhodocyclaceae* (48% and 22%). The second most abundant groups in this gate belonged to unclassified *Proteobacteria* (16 and 8%), followed by the family *Brucellaceae* (8%). The families *Caulobacteriaceae, Xanthomonadaceae,* and the phylum *Bacteroidetes* constituted a small proportion (15%) of the total community in the gate. The role of the identified bacterial groups in the degradation of the retanning agents is reflected by the positive correlation ($p < 0.05$) detected between cell abundances and removal of the retanning agents (**Figure 9**).

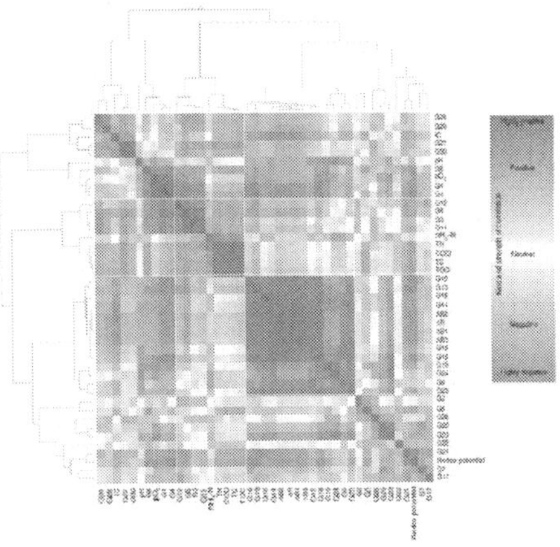

Figure 8. Correlation of 13 abiotic parameters with cell abundances in the 30 gates during the running period of the reactor (after Ref. [36]).

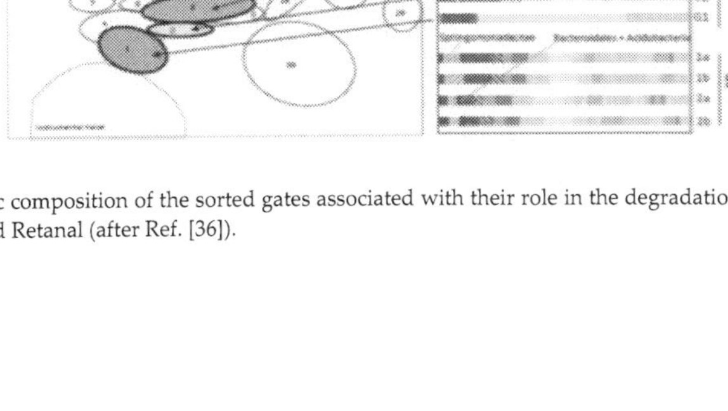

Figure 9. Taxonomic composition of the sorted gates associated with their role in the degradation of the two retanning agents Basyntan and Retanal (after Ref. [36]).

4. Conclusion

The findings of this study provided a preliminary investigation on the biodegradability of two of the several types of xenobiotic compounds used in the tanning industry. It was possible to single out bacterial groups such as *Bacteroidetes* and *Proteobacteria,* with strong correlation with the complete degradation of some of the compounds in retanning chemicals.

Management of wastewater treatment plants (WWTPs) primarily focuses on process parameters and physicochemical (abiotic) properties of the wastewater before and after treatment. Stable performance of any biological wastewater treatment system can be achieved by understanding and manipulating the microbial communities residing in the system besides the management of the conventional process parameters and abiotic properties. Investigations of microorganisms responsible for efficient reduction of pollutants in various biological wastewater treatment plants have been conducted for many years. This study was successful in identifying bacterial groups involved in different nutrient removal processes from tannery wastewater such as sulfur oxidation, denitrification, and cyclic aromatic compound degradation. Moreover, this study is one of the few studies conducted in field-scale reactors that integrate different approaches to interpret the functional property of a biological treatment system.

Acknowledgements

The authors like to acknowledge the Swedish International Development Cooperation Agency (SIDA), Addis Ababa University, and BIO-EARN (East African regional program and research network working for biotechnology, biosafety, and biotechnology policy development) for the financial support on the establishment of the WWTP. Biosciences eastern and central Africa, International Livestock Research Institute (BecA-ILRI Hub) and the German Academic Exchange Service (DAAD) are highly acknowledged for the financial and technical support of the various analyses of this study.

Author details

Adey Feleke Desta[1,*], Joyce Nzioki[2], Solomon Maina[3] and Francesca Stomeo[2]

*Address all correspondence to: adey.desta@gmail.com

1 Institute of Biotechnology, Addis Ababa University, Addis Ababa, Ethiopia

2 Biosciences eastern and central Africa-International Livestock Research Institute (BecA-ILRI) Hub, Nairobi, Kenya

3 School of Plant Biology, University of Western Australia, Perth, Western Australia, Australia

References

[1] COTANCE (Confederation of Tanning Industries of the European Union). The European Tanning Industry Sustainability Review. Brussels, Belgium; 2002. 96pp.

[2] Heidemann E. Fundamentals of Leather Manufacturing. Eduard Roether KG: Darmstadt, Germany; 1993. 649 p.

[3] Sharphouse HJ. Leather Technician's Handbook. Vernon Lock Ltd; London, 1983. 52 p.

[4] Khan SR, Khwaja MA and Khan AM. Environmental impacts and mitigation costs associated with cloth and leather exports from Pakistan. Environ. Dev. Econ. 2001; **6**: 383-403.

[5] Jankauskaite V, Gulbiniene A, Urbelis V, Jiyembetova I, Sirvaityte J and Beleska K. Comparable evaluation of leather waterproofing behaviour upon hide quality. I. Influence of retanning and fatliqouring agents on leather structure and properties. Mater. Sci. (Medziagotyra). 2012; **18**: 150-157.

[6] Anilin B and Fabrik S. Pocketbook for the Leather Technologist. 4th ed. Ludwigshafen: BASF; 2007. 454 pp.

[7] Naviglio B, Calvanese G, Tortora G, Cipollaro L and Pierri G. Characterization of Tannery Chemicals: Retanning Agents. Naples, Italy: Stazione Sperimentale Per L'industria Delle Pelli E Delle Materie Concianti; 1996. 20 p.

[8] Piotte M, Bossanyi F, Perreault F and Jolicoeur C. Characterization of poly-naphthalenesulfonate salts by ion-pair chromatography and ultrafiltration. J. Chromatogr. A 1995; **704**: 377-385.

[9] Bienkiewicz K. Physical Chemistry of Leather Making. Malabar, Florida: Krieger Publishing Co.; 1983. 541 p.

[10] Song Z, Suzanne R, Howland KE and Burns RG. Analysis of a retan agent used in the tanning process and its determination in tannery wastewater. Anal. Chem. 2003; **75**: 1285-1293.

[11] Ludvick J, editor. The scope for decreasing pollution load in leather processing. Regional program for pollution control in the tanning industry in South East Asia. UNIDO/US/RAS/92/120. 2000. 51 p.

[12] Teodorescu M and Gaidau C. Possible steps to follow for filling the gap between requirements and the real condition in tanneries. J. Cleaner Prod. 2007; **16**: 622-631.

[13] Alexander KTW, Corning DR, Cory NJ, Donohue VJ and Sykes RL. Environmental and safety issues—clean technology and environmental auditing. J. Soc. Leath. Tech. 1991; **76**: 17-23.

[14] Calheiros CSC, Rangel AOSS, Castro PML. Constructed wetland systems vegetated with different plants applied to the treatment of tannery wastewater. Water Res. 2007; **41**: 1790-1798.

[15] Apaydin O, Kurt U and Gonullu MT. An investigation on the tannery wastewater by electrocoagulation. Global NEST J. 2009; **11**: 546-555.

[16] Ganesh R, Balaji G and Ramaunjam RA. Biodegradation of tannery wastewater using sequencing batch reactor—respirometric assessment. Bioresour. Technol. 2006; **97**: 1815-1821.

[17] Haydar S, Aziz JA and Ahmad MS. Biological treatment of tannery wastewater using activated sludge process. Pakistan J. Eng. Appl. Sci. 2007; **1**: 61-66.

[18] Kongjao S, Damronglerd S and Hunsom M. Simultaneous removal of organic and inorganic pollutants in tannery wastewater using electrocoagulation technique. Korean J. Chem. Eng. 2008; **25**: 703-709.

[19] Koteswari YN and Ramaniba R. The effect of tannery effluent on the colonization rate of plankets: a microcosm study. Turk. J. Biol. 2003; **27**: 163-170.

[20] Lefebvre O, Vasudevan N, Torrijos M, Thanasekaran K and Moletta R. Halophilic biological treatment of tannery soaks liquor in a sequencing batch reactor. Water Res. 2005; **39**: 1471-1480.

[21] Leta S. Developing and optimizing processes for biological nitrogen removal from tannery wastewaters in Ethiopia [thesis]. Stockholm, Sweden: Royal Institute of Technology (KTH); 2006. ISBN 91-7283-830-2. 77pp.

[22] Orhon D, Ates E and Sözen S. Experimental evaluation of the nitrification kinetics for tannery wastewaters. Water S. A. 2000; **26**: 43-52.

[23] Ram B, Bajpai PK and Parwana HK. Kinetics of chrome-tannery effluent treatment by the activated sludge system. Process Biochem. 1999; **35**: 255-265.

[24] Szpyrkowicz L, Kaul SN and Neti RN. Tannery wastewater treatment by electro-oxidation coupled with a biological process. J. Appl. Electrochem. 2005; **35**: 381-390.

[25] Thanigavel M. Biodegradation of tannery effluent in fluidized bed bioreactor with low density biomass support [thesis]. Tamil Nadu, India: Annamalai University; 2004.

[26] Reemtsma T, Fiehn O, Kalnowski G and Jekel M. Microbial transformations and biological effects of fungicide-derived benzothiazoles determined in industrial wastewater. Environ. Sci. Technol. 1995; **29**: 478-485.

[27] Fiehn O, Wegener G and Jekel M. Capillary electrophoretic analysis of organic anions in tannery wastewater including high contents of chloride and sulfate. Int. J. Environ. Anal. Chem. 1994; **69**: 257-271.

[28] UNIDO (United Nations Industrial Development Organization). Introduction to treatment of tannery effluents: What every tanner should know about effluent treatment. Vienna: UN Publication; 2011.

[29] APHA (American Public Health Association). Standard Methods for the Examination of Water and Wastewater: 20th ed. Washington, DC, USA: American Public Health Association/American Water Works Association/Water Environment Federation; 1998.

[30] Andrews S. FastQC a quality control tool for high throughput sequence data [internet]. 2010. Available at: http://www.bioinformatics.babraham.ac.uk/projects/fastqc/ [Accessed 2014-09-21]

[31] Hannon LabFastX Toolkit [internet]. 2010. Available at: http://hannonlab.cshl.edu/fastx_toolkit [Accessed 2014-09-21]

[32] Zerbino DR and Birney E. Velvet: algorithms for de novo short read assembly using de Bruijn graphs. Genome Res. 2008; **18**: 821-829. doi:10.1101/gr.074492.107

[33] Haas BJ, Papanicolaou A, Yassour M, Grabherr M, Blood PD, Bowden J, Couger MB, Eccles D, Li B, Lieber M, Macmanes MD, Ott M, Orvis J, Pochet N, Strozzi F, Weeks N, Westerman R, William T, Dewey CN, Henschel R, Leduc RD, Friedman N, Regev A. De novo transcript sequence reconstruction from RNA-seq using the Trinity platform for reference generation and analysis. Nature Protocols. 2013; **8**:1494-512. doi: 10.1038/nprot.2013.084

[34] Schmieder R, Lim YW and Edwards R. Identification and removal of ribosomal RNA sequences from metatranscriptomes. Bioinformatics. 2012; **28**:433-435.

[35] Conesa A, Götz S, García-Gómez JM, Terol J, Talón M and Robles M. Blast2GO: a universal tool for annotation, visualization and analysis in functional genomics research. Bioinformatics 2005; **21**: 3674-3676. doi:10.1093/bioinformatics/bti610

[36] Koch C, Guenther S, Desta AF, Huebschmann T, Mueller S.Cytometric fingerprinting for analyzing microbial intracommunity structure variation and identifying sub-community function. Nature Protocols 2013; **8**: 190-212. doi:10.1038/nprot.2012.149

[37] Desta AF, Assefa F, Leta S, Stomeo F, Wamalwa M, Njahira M. Microbial community structure and diversity in an integrated system of anaerobic-aerobic reactors and a constructed wetland for the treatment of tannery wastewater in Modjo, Ethiopia. PLoS One. 2014; **9**: e115576. doi: 10.1371/journal.pone.0115576. pmid:25541981

Section 3

Biofilm for Wastewater Treatment

Chapter 9

Application of Mixed Microbial Culture Biofilms for Manganese (II), Cobalt (II), and Chromium (VI) Biosorption by Horizontal Rotating Tubular Bioreactor

Tonči Rezić, Iva Rezić, Michaela Zeiner and
Božidar Šantek

Additional information is available at the end of the chapter

http://dx.doi.org/10.5772/66920

Abstract

Industrial wastewater contaminated with toxic heavy metals is a big ecological and environmental problem. Applying biological materials to effectively remove and recover heavy metals from contaminated wastewaters has gained importance as promising alternative to conventional treatment techniques. Thus, the objective of the presented paper is the investigation of the capability of microorganisms, isolated from polluted (metal-laden) soil, to biosorb toxic metals from aqueous solutions. Biosorption process for heavy metal removal was conducted in a new pilot scale horizontal rotating tubular bioreactor (HRTB). This bioreactor provides conditions for microorganism's growth in a form of suspended cells and biofilm. Biofilm is capable to protect microorganisms from interaction with toxic metals in the surrounding environment. Three metals were selected as model examples: cations of manganese and cobalt and hexavalent chromium (an oxyanion). Optimized bioreactor conditions, namely, medium inflow rate (F) and bioreactor rotation speed (n) for biofilm formation and metal removal were monitored, and under optimized bioreactor conditions, promising results were obtained.

Keywords: heavy metals, mixed microbial culture, biosorption, biofilm, horizontal rotating tubular bioreactor

1. Introduction

Heavy metal's wastewater pollution has always been a very serious problem because these elements are not biodegradable and can accumulate in living tissues causing serious health effects [1]. Heavy metals are introduced into the natural environment through many industrial processes including leather tanning, wood preservation, metal plating, mining operations, chloralkali, radiator manufacturing, smelting, alloy industries, storage batteries, and automobile manufacturing [2]. Since the early 1970s, there has been growing concern over the effect of heavy metals on humans and aquatic ecosystems [3]. The Agency for Toxic Substances and Disease Registry (ATSDR) classifies nickel as a human carcinogen based on its chronic and subchronic effects [4]. Iron and copper can cause stomach and intestinal distress, liver and kidney damage, and anemia. Zinc may cause anemia, damage the pancreas, and decrease levels of high-density lipoprotein (HDL) cholesterol [5].

There are many conventional methods (physical and chemical) for heavy metal removal, but in general, they result with much waste which is hard to treat. In addition, several widely applied processes such as ion exchange, membrane technologies, and adsorption are very expensive processes when used for large quantities of wastewater which contain low concentrations of heavy metals [6]. Heavy metals may be removed from water as an insoluble soil by the chemical process of precipitation, respectively. However, chemical treatment of heavy metals generates environmentally hazardous chemical byproducts. Additionally, chemical treatment requires constant adjustment of pH value to a narrow range for optimal heavy metal removal, thereby increasing the labor input and cost [7].

As an alternative, different biochemical methods can be applied because they do not destroy metals, but concentrate and immobilize them [8]. Biosorption is removal of metals and their complexes from samples by biological materials [9]. Bioadsorbens can efficiently remove heavy metals from solutions with low concentration; therefore, they are ideal adsorptive media for wastewaters with low concentrations of metal ions. Microbial metal accumulation has received much attention during recent years, due to the potential use of microorganisms for treatment of metal-polluted water or wastewater streams. Recently, several bacterial species have been identified to remove toxic heavy metals [10, 11]. Biosorption can be performed on live or dead microorganisms, as well as on their parts or extracellular products and microorganism aggregations on the surfaces in the structures called biofilm. Biofilm application in the biosorption showed great potential in the wastewater treatment systems. Different types of bioreactor systems such as trickling filters, fluidized or packed bed bioreactors, and thin layer or biodisc reactors were implemented for biofilm formation and wastewater treatment [12–14]. Horizontal rotating tubular bioreactor (HRTB) was designed as combination of a thin layer [15, 16] and biodisc reactor [17] with construction abilities for successful biofilm formation. Consequently, bioreactor interior is equipped with o-shaped partitional walls which provide area for biofilm formation. Wide investigation of HRTB mixing properties was previously done [18–22], and aerobic and anaerobic bioprocesses were successfully conducted. As a model of anaerobic bioprocess, fermentative glucose conversion was chosen [23]. Acetate removal with mixed microbial culture was selected as a model bioprocess for study of HRTB performance in aerobic condition [24]. As combination of aerobic and anaerobic bioprocesses, nitrification and denitrification were done in two consecutive steps in the same bioreactor vessel [25].

In this investigation HRTB was used for native mixed microbial biofilm formation and investigation of developed biofilm biosorption abilities. In the biosorption experiments, artificial wastewater with heavy metal ions Co(II), Cr(VI), and Mn(II) was applied as representative example of textile industry wastewaters. Observed results showed significant potential of developed mixed microbial culture biofilm to successfully remove toxic heavy metals in applied bioreactor.

2. Material and methods

2.1. Microorganism, medium, and growth conditions

Mixed microbial culture was isolated from surface sediments sampled in the Kaštela bay—industrial area located near town Split, at the Croatia Adriatic coast. Isolation was done from 5 g of soil samples. Samples were resuspended in Erlenmeyer flasks with different contents of heavy metals in feeding medium (**Table 1**) and cultivated 48 h at 23 ± 1°C. Rotation speed during cultivation was 150 rpm. After 48 h flat plates were inoculated with 1 mL sample from each flask. Medium content used for flat plate cultivation was the same as shown in **Table 1** with 20 g/L of agar. Viable cells were determined as colony-forming units (CFU 1/mL). The number of colonies was counted after 48 h at 23 ± 1°C. Only medium 1 provided satisfied condition for microorganism colony forming. Therefore, this medium was used for cultivation in tank bioreactor and HRTB. In this research, the medium was sterilized at 121°C for 20 min.

Content (g L^{-1})	Medium 1	Medium 2	Medium 3	Medium 4
Glucose	10.00	10.00	10.00	10.00
Yeast extract	3.00	3.00	3.00	3.00
Tripton	3.00	3.00	3.00	3.00
$CuSO_4{\cdot}5H_2O$	–	0.49	0.49	0.49
$ZnSO_4{\cdot}7H_2O$	–	0.55	–	0.55
$CoCl_2{\cdot}6H_2O$	0.51	0.50	0.50	0.50
$FeSO_4{\cdot}7H_2O$	–	0.62	0.62	0.62
$MnSO_4{\cdot}H_2O$	0.39	–	0.39	0.39
$NiSO_4{\cdot}6H_2O$	0.56	–	0.56	0.56
$K_2Cr_2O_7$	0.20	0.20	–	0.20
$C_4H_6O_4Pb{\cdot}3H_2O$	–	–	0.23	0.23

Table 1. Contents of feeding medium used during microorganism isolation from soil samples.

2.2. Characteristics and experimental setup of the bioreactor

The HRTB is a stainless steel tube with 2.0 m length and 0.25 m diameter. O-ring–shaped partition walls (inner diameter 0.19 m) divide its interior in a 0.02 m long section. The liquid volume of the bioreactor was 15 L. In order to enable rotation of the entire reactor, the HRTB is horizontally placed on appropriate bearings. The aeration was performed via the central tube fixed in the bioreactor's axis. Improvement of the aeration was obtained by submerging the aeration tube on five positions along the HRTB. For all experimental works, the airflow rate was 152 L h^{-1}. In **Figure 1** the sampling systems for broth and biofilm are shown, being places at

0.40 m intervals. On the cover of the sampling place, a flat plate (0.02 × 0.02 m) is fixed as device for biofilm thickness measurement.

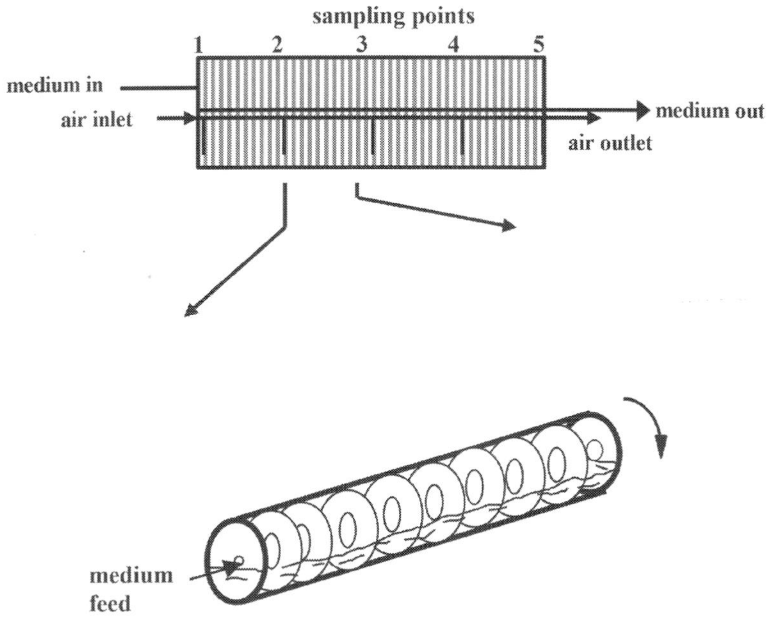

Figure 1. Schematic diagram of HRTB and the inner structure of HRTB with O-ring-shaped partition walls.

Batch cultivation in a stirred tank bioreactor is used to obtain the suspended bacterial biomass (7.5 L) needed for inoculating the HRTB. The feeding process was started after 24 h at a rate of 1 L h^{-1} and a rotation speed of the HRTB of 10 min^{-1}. A stable biofilm in the bioreactor is available after 15 days, which is considered as ready to start the experiments with different parameter variations, such as medium inflow rate (0.5, 1.0, and 2.0 L h^{-1}) and bioreactor rotation speed (5, 15, and 30 min^{-1}). The dynamics of the bioprocess in HRTB was monitored by withdrawing the samples from five positions along the bioreactor length after five residence times since the new set of process parameters was established. The bioreactor was operated under a constant influent glucose concentration 10 g L^{-1} and metal ion concentration Co^{2+} = 0.125 g L^{-1}, Mn^{2+} = 0.125 g L^{-1}, and Cr^{6+} = 0.125 g L^{-1}.

Since it was known in previous studies that bioreactor rotation speed higher than 30 min^{-1} leads to intensive biofilm detachment [23, 25], no higher speed are tested in the current investigation. Experiments with varying bioreactor rotation speed are carried out prior to changes of medium inflow rate, since the latter have exhibited higher effects on the bioprocess dynamics and the biofilm stability [23, 24, 26, 27].

2.3. Analytical methods

Biomass concentration in suspension was determined by centrifuging the culture medium of 35 mL for 20 min at 4500 rpm (3629 g), washing twice with demineralized water and then drying at 105°C/48 h. Supernatants were used for determination of Co^{2+}, Mn^{2+}, and Cr^{6+} (UV-Vis spectrophotometrical method by Fries and Getrost) [28]. All determinations were done in triplicates.

Inductively coupled plasma-mass spectrometry (ICP-MS) was used to quantify the metals in biofilm and suspended biomass after acidic digestion. The spectrometer used had a GemCone nebulizer, a cyclone spray chamber, and a standard one-piece extended torch with a quartz injector tube. Each metal was quantified by measurements in triplicates at three different wavelengths. The biofilm samples were mineralized using a closed microwave digestion system. Each sample was digested with a mixture of 5 mL nitric acid, 1 mL hydrogen peroxide, and 1 mL double-distilled water. The digestion was performed in five steps—3 min at 250 W, 1 min without power, 4.5 min at 250 W, 6 min at 650 W, and 5 min at 400 W—followed by a ventilation time of 25 min.

The biofilm thickness was measured applying a modified Venkataraman and Ramanujam method [29]: graphite powder was used instead of chalk powder. The projector was replaced by a microscope with micrometric scale. In order to determine the mass of the biofilm, samples were collected from the inner surface of HRTB, suspended in demineralized water, and twice washed after centrifugation. Finally the biofilm samples were dried for 48 h at 105°C.

Suspended biomass sorption capacity ($q_{x,L}$) was calculated as follows:

$$q_{x,L} = \frac{m_M}{m_{x,L}} \tag{1}$$

where m_M mass of metal ion (mg) and $m_{x.L}$ is the dry weight of suspended biomass (g).

2.4. Mathematical model development

2.4.1. *Diffusion process*

The diffusivity of metal ion in water was estimated using the Wilke-Chang equation [30]:

$$D_{aq} = 7.4 \cdot 10^{-8} \frac{\left(\xi_{aq} M_{aq}\right)^{1/2} T}{V_M{}^{0.6} \eta_{aq}} \tag{2}$$

where D_{aq} is the diffusion coefficient of metal ion in water ($m^2\ s^{-1}$), ξ is the metal ion connecting factor, V_M is the metal ion molar volume ($m^3\ mol^{-1}$), η is the water dynamic viscosity ($kg\ m^{-1}\ s^{-1}$), T is the temperature (K), and M_{aq} is the water molecular mass ($kg\ mol^{-1}$).

Metal ion relative diffusivity (f_D) was computed from Horn-Morgenroth equation [31]:

$$f_D = 1 - \frac{0.43 c_{x,f}^{0.92}}{11.19 + 0.27 c_{x,f}^{0.99}} \tag{3}$$

where $c_{x,f}$ is the biofilm density (kg m^{-3}).

Effective diffusion coefficient of metal ion in biofilm was calculated using this correlation [32]:

$$D_{ef,M} = f_D D_{aq} \tag{4}$$

where $D_{ef,M}$ is the effective diffusion coefficient of metal ion in biofilm ($m^2\ s^{-1}$).

Mass transport of all dissolved metal ions in biofilm follows Fick's second law of molecular diffusion:

$$D_{ef,M} \frac{\partial^2 c_{M,f}}{\partial z^2} = \frac{\partial c_{M,f}}{\partial t} \tag{5}$$

where $c_{M,f}$ is the concentration of metal ion in biofilm phase (kg m^{-3}), t is time (s), and z is biofilm depth (m).

2.4.2. *One-dimensional diffusion-bioadsorption model*

In the dynamic equilibrium conditions, metal ion concentration in the biofilm is represented conceptually as functions of biofilm depth z as shown in **Figure 2C**. Concentrations of metal ion ($c_{M,f}$) in biofilm phase are given by the second-order polynomial correlation:

$$c_{M,f}(z) = a_0 + a_1 z + a_2 z^2 \tag{6}$$

where a_0, a_1, and a_2 are the second-order polynomial correlation coefficient and z is biofilm depth coordinate (m).

Metal ion concentrations in the bulk liquid phase ($c_{M,L}$) represent as constant values for each ideal mixing segment (**Figure 2B**, **C**).

The biofilm zone is surrounded by the stagnant liquid layer of thickness L_g (**Figure 2C**). The mass transfer coefficient (k_m) in the stagnant liquid layer was estimated by the correlation [18, 19]:

$$k_m = 0.664\,(D_{tb}/L_k)\, Re_N^{1/2} Sc^{1/3} \tag{7}$$

where L_K is the wetted perimeter of bioreactor (0.254 m), D_{tb} = D_{aq} is the diffusion coefficients of metal ions in water ($m^2\ h^{-1}$), Sc is Schmidt number, and Re_N is Reynolds rotation number. Schmidt number (Sc) was calculated from [33]

$$Sc = \nu / D_{tb} \tag{8}$$

where ν is kinematic viscosity ($m^2\ s^{-1}$).

Reynolds rotation number (Re_N) of HRTB was calculated by following equation [18]:

$$\mathrm{Re}_N = \frac{D_{UP}\pi n L_K \rho}{2\eta} + \frac{D_T \Pi n L_K \rho}{2\eta} \tag{9}$$

where D_{UP} is the inner diameter of partition wall in HRTB (m), L_K is wetted perimeter of bioreactor (0.254 m), Π is the Ludolph's number (3.14159), D_T is the bioreactor diameter (m), n is the bioreactor rotation speed (s^{-1}), ρ is the broth density (kg m^{-3}), and η is the dynamic viscosity of broth (kg m^{-1} s^{-1}).

Regarding to "spiral flow" mixing model [18, 19], based on the physical model which divided the bioreactor into ideally mixed compartments (**Figure 2A**), mass balances of the heavy metal ion for the first ideal mixing segment across the bulk liquid (**Figure 2A, B**) were

$$V_L^{1,1}\frac{dc_{M,L}^{1,1}}{dt} = F_u c_{M,L}^0 + F_{cr}c_{M,L}^{1,Ni} + F_p c_{M,L}^{2,1} - (F_u + F_p)c_{M,L}^{1,1} - F_{cr}c_{M,L}^{1,1} - V_L^{1,1}r_{M,L}^{1,1} \tag{10}$$

where $c_{M,L}^{1,1}$ is liquid section metal ion concentrations in the first segment (Ni = 1) of the first kaskade (Nl = 1) (kg m^{-3}), $c_{M,L}^0$ is inflow metal ion concentration (kg m^{-3}), F_u is inflow (m^3 h^{-1}), F_p is back flow (m^3 h^{-1}), F_{cr} is circulation flow (m^3 h^{-1}), ${r_{M,L}}^{1,1}$ is liquid section reaction rate in the first segment (Ni = 1) of first kaskade (Nl = 1) (kg m^{-3} h^{-1}), ${V_L}^{1,1}$ is liquid section volume in the first segment (Ni = 1) of the first kaskade (Nl = 1) (m^3), and $c_{M,L}^{1,Ni}$ is liquid section metal ion concentrations in the Ni-segment of the first kaskade (kg m^{-3}).

First ideal mixing segments of all cascades were represented in the model without biofilm zone (**Figure 2B**). All other ideal mixing segments include biofilm zone (**Figure 2C**). Therefore, mass

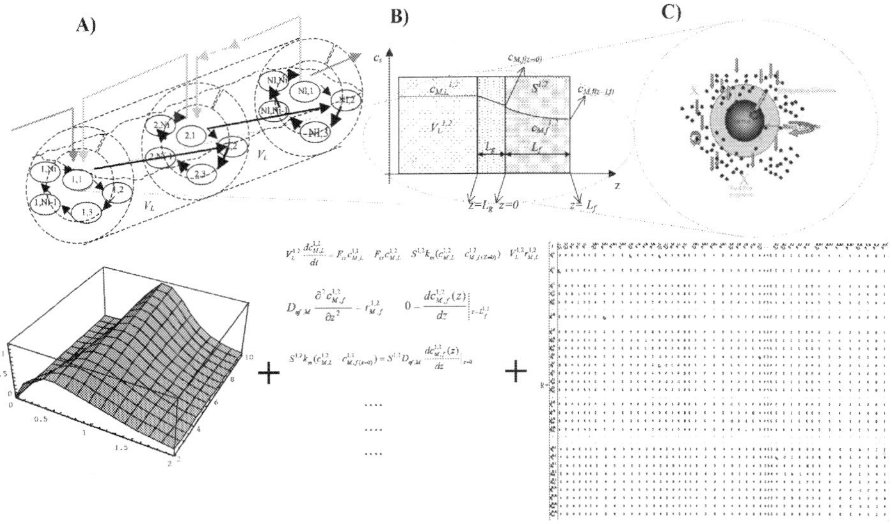

Figure 2. Conceptual representation of metal biosorption in HRTB: (A) "spiral flow" mixing, (B) metal ion diffusion, and (C) biosorption reaction.

balances of the heavy metal ion were computed across the bulk liquid for the second ideal mixing segment (**Figure 2A, C**) as follows:

$$V_L^{1,2}\frac{dc_{M,L}^{1,2}}{dt} = F_{cr}c_{M,L}^{1,1} - F_{cr}c_{M,L}^{1,2} - S^{1,2}k_m\left(c_{M,L}^{1,2} - c_{M,f(Z=0)}^{1,2}\right) - V_L^{1,2}r_{M,L}^{1,2} \tag{11}$$

where $c_{M,L}^{1,2}$ is liquid section metal ion concentrations in the second segment ($Ni = 2$) of the first kaskade ($Nl = 1$) (kg m^{-3}), $c_{M,f(Z=0)}^{1,2}$ is metal ion concentration in the second segment ($Ni = 2$) of the first kaskade ($Nl = 1$) on the biofilm surface (kg m^{-3}), $V_L{}^{1,2}$ is liquid section volume in the second segment ($Ni = 2$) of the first kaskade ($Nl = 1$) (m^3), $S^{1,2}$ is mass transfer surface in second ideal mixing segment ($Ni = 2$) of the first kaskade ($Nl = 1$) (m^2), and $r_{M,L}{}^{1,2}$ is liquid section reaction rate in the second segment ($Ni = 2$) of the first kaskade ($Nl = 1$) (kg m^{-3} h^{-1}).

Liquid section volume in the ideal mixing segment ($V_L{}^{Nl,Ni}$) was computed from the bioreactor liquid volume using the following equation:

$$V_L^{Nl,Ni} = \frac{V_L}{Nl \cdot Ni} \tag{12}$$

where $V_L{}^{Nl,Ni}$ is liquid section volume in the ideal mixing segment (m^3), V_L is liquid volume in the HRTB (m^3), Nl is the number of kaskades, and Ni is the number of ideal mixing segments.

Mass transfer surface in the ideal mixing segment ($S^{Nl,Ni}$) was computed from the inside bioreactor surface using the following equation:

$$S^{Nl,Ni} = \frac{S}{Nl \cdot Ni} \tag{13}$$

where $S^{Nl,Ni}$ is mass transfer surface in the ideal mixing segment (m^2) and S is inside bioreactor surface (m^2).

Mass transport of all dissolved metal ions in biofilm is derived from Eq. (5) and equal to reaction rate ($r_{M,f}^{1,2}$):

$$D_{ef,M}\frac{\partial^2 c_{M,f}^{1,2}}{\partial z^2} = r_{M,f}^{1,2} \tag{14}$$

The inner boundary conditions (at $z = 0$) at biofilm-liquid interface are given as

$$S^{1,2}k_m\left(c_{M,L}^{1,2} - c_{M,f(z=0)}^{1,2}\right) = S^{1,2}D_{ef,M}\frac{dc_{M,f}^{1,2}(z)}{dz}\Big|_{z=0} \tag{15}$$

The outer boundary conditions (at $z = L_f{}^{1,2}$) at biofilm-bioreactor interface are given as

$$0 = \frac{dc_{M,f}^{1,2}(z)}{dz}\Big|_{z=L_f^{1,2}} \tag{16}$$

As mentioned before, concentrations of the metal ion in the biofilm are represented with second-order polynomial correlation [Eq. (6)]. Assuming dynamic equilibrium conditions at time (t) model were derived from mass balances equation [Eqs. (11), (14), (15)] and second-order polynomial correlation for metal ion concentration [Eq. (6)], taken across biofilm zone vertical to the biofilm surface [Eqs. (17)–(20) below]:

Bulk liquid section:

$$0 = F_{cr}c_{M,L}^{1,1} - F_{cr}c_{M,L}^{1,2} - S^{1,2}k_m\left(c_{M,L}^{1,2} - a_0^{1,2}\right) - V_L^{1,2}r_{M,L}^{1,2} \tag{17}$$

Biofilm zone:

$$D_{ef,M}2a_2^{1,2} = r_{M,f}^{1,2} \tag{18}$$

The inner boundary conditions (at $z = 0$):

$$-\frac{k_m}{D_{ef,M}}\left(c_{M,L}^{1,2} - a_0^{1,2}\right) = a_1^{1,2} \tag{19}$$

The outer boundary conditions (at $z = L_f^{1,2}$):

$$a_1^{1,2} = -2a_2^{1,2}L_f^{1,2} \tag{20}$$

where $a_0^{1,2}$, $a_1^{1,2}$, and $a_2^{1,2}$ are the second-order polynomial correlation coefficient in the second segment ($Ni = 2$) of the first kaskade ($Nl = 1$); $L_f^{1,2}$ biofilm thickness in the second segment ($Ni = 2$) of the first kaskade ($Nl = 1$) (m); $V_L{}^{1,2}$ is liquid section volume in the second segment ($Ni = 2$) of the first kaskade ($Nl = 1$) (m^3); and $r_{M,L}{}^{1,2}$ is liquid section reaction rate in the second segment ($Ni = 2$) of the first kaskade ($Nl = 1$) (kg m^{-3} h^{-1}).

Adjusting mass balances and reaction rates for all ideal mixing segments according to **Figure 2**, system of the differential equations was developed for heavy metal ion concentration changes along HRTB.

2.4.3. *Bioadsorption kinetic model*

Mass balance equations were coupled to the reaction rate terms in the liquid section ($r_{M,L}$) and in the biofilm zone ($r_{M,f}$) based on the Freundlich adsorption isotherm. Instead kinetic terms heavy metal removal was changed with bioadsorption model [34] [Eqs. (21), (22)]:

$$q_{x,L} = K_F(c_{M,L})^{1/h} \tag{21}$$

$$q_{x,f} = K_F\left(c_{M,f}\right)^{1/h} \tag{22}$$

were $q_{x,L}$ is suspended biomass adsorption capacity (mg g^{-1}), $q_{x,f}$ is biofilm adsorption capacity (mg g^{-1}), and K_F and h are Freundlich isotherm constant.

Bioadsorption model for biofilm zone was derived from Freundlich equation [Eq. (22)] and second-order polynomial correlation for metal ion biofilm concentration [Eq. (6)]:

$$q_{x,f} = K_F \left[a_0 + a_1 z_i + a_2 (z_i)^2 \right]^{1/h} \tag{23}$$

where z_i is collocation point across biofilm zone parallel to the substratum surface.

Assuming the collocation point $z_i = L_f/b$ where $b \in N(+)$ bioadsorption model are defined as follows:

$$q_{x,f} = K_F \left[a_0 + a_1 \frac{L_f}{b} + a_2 \left(\frac{L_f}{b} \right)^2 \right]^{1/h} \tag{24}$$

The kinetic model assumes that reaction rate is the function of biomass concentration in the liquid section ($c_{x,L}$) and in the biofilm zone ($c_{x,f}$) [Eqs. (25), (26) below]:

$$r_{M,L} = \frac{c_{x,L} q_{x,L}}{\tau} \tag{25}$$

$$r_{M,f} = \frac{c_{x,f} q_{x,f}}{\tau} \tag{26}$$

where $r_{M,f}$ is biofilm section reaction rate (kg m^{-3} h^{-1}), $r_{M,L}$ is liquid section reaction rate (kg m^{-3} h^{-1}), and τ is retention time (h).

2.4.4. Numerical methods

The model equations were solved by personal computer using the "Wolfram Mathematica" program routine "NDSolve, FindRoot, FindMinimum, Fit," and orthogonal collocation methods [35–37] were applied for the inner biofilm concentration profiles representing.

2.4.5. Initial parameter values

The model was initially simulated using kinetic parameters (K_F and h) from previous studies [38] and mixing parameters (Nl, Ni, F_{cr}, and F_p) computed in this study (**Table 1**). Transport parameters include the mass transfer coefficient rate of metal ions (k_m), and the effective diffusion coefficient of metal ion in biofilm ($D_{ef,M}$) was estimated by Eqs. (7) and (2)–(4).

2.4.6. Parameter optimization

The empirical equations developed from HRTB mixing modeling were used as a fitness function during mixing parameter optimization (Nl, Ni, F_{cr}, and F_p). Kinetic parameters (K_F and h) were optimized computing variance between observed variables and simulated variables as

$$E_n = \frac{1}{n_u} \sum_{i=1}^{i=n_u} \frac{\left(c^i_{n,\exp} - c^i_{n,sim} \right)^2}{c^i_{n,\exp}} \tag{27}$$

where $c^i_{n,\exp}$ is observed variables (kg m^{-3}), $c^i_{n,sim}$ is simulated variables (kg m^{-3}), and n_u is number of observations.

To determine dependence of parameter change on variance between observed variables and simulated variables (E_n), calculation were performed by polynomial regression with the "Wolfram Mathematica" routine "Fit." After this plug, optimization was preformed calculating global minimum variance between observed variables and simulated variables using routine "FindMinimum."

3. Results and discussion

3.1. Biofilm formation studies in HRTB

In this work the effect of process parameters (n and F) on the mixed microbial culture biofilm formation in HRTB was studied as a continuation of comprehensive research of mixing [18–20] and conduction of model bioprocesses in HRTB [23–27]. This investigation started with mixed microbial culture isolation from surface sediments highly contaminated with heavy metals [39–41].

Isolated mixed microbial culture was developed in HRTB as described in Section 2.2, whereby the culture first grew in suspension and then a biofilm was gradually established on the O-shaped rings and inner surface of bioreactor. **Figure 3** represents O-shaped rings before and after biofilm formation.

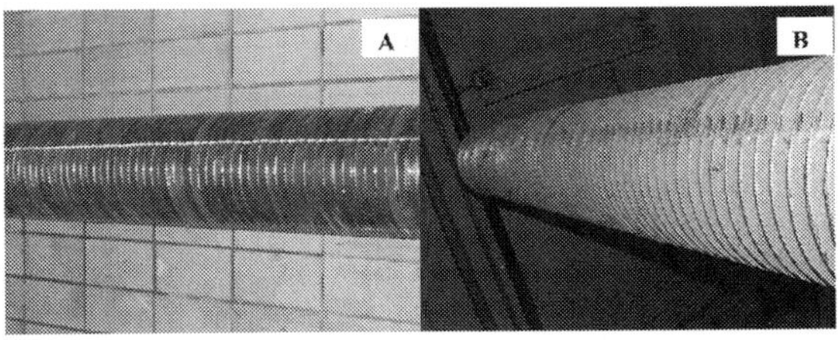

Figure 3. O-shaped rings before (A) and after (B) biofilm formation in HRTB.

The biofilm obtained was used for the investigation of suspended biomass adsorption abilities and biofilm properties (thickness, density) by different combinations of process parameters. Changes of process parameters (n and F) during this investigation are presented in **Figure 4**.

A significant disturbance was observed at F = 2.0 L h^{-1} and n = 30 min^{-1} when biofilm detachment took place. Influence of biofilm detachment on suspended biomass concentration changes will be discussed in the next section.

3.2. Suspended biomass concentration and biosorption capacity in HRTB

In the present study, biomass grew as suspended single cells, suspended cell clusters, and biofilm attached to the bioreactor inner surface. **Table 2** shows the results of suspended biomass concentration in dependency of parameter variation: inflow rate (F = 0.5–2.0 L h^{-1}) and bioreactor rotation speed (n = 5–30 min^{-1}). The suspended biomass concentrations ($c_{x.L}$) range from 0.95 to 1.07 g L^{-1} at inflow rate 0.5 L h^{-1}. The increase of the inflow rate to 1.0 and

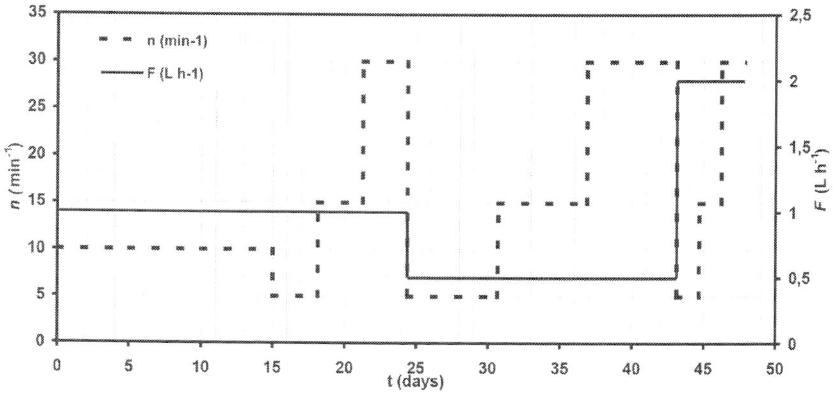

Figure 4. Dynamics of process parameter changes [bioreactor rotation speed (n) and medium inflow rate (F)] during investigation in the HRTB.

2.0 L h^{-1} was related to the increase of suspended biomass concentrations (1.59–5.11 g L^{-1}) as a consequence of biofilm detachment and erosion. Highest suspended biomass concentration was 5.11 g L^{-1} registrated as a consequence of more intensive biofilm detachment (release of larger biofilm parts) due to high inflow rate (F = 2.0 L h^{-1}) and bioreactor rotation speed (n = 30 min^{-1}). In this situation, considerable increase of metal ion concentrations was observed as a consequence of biomass washout from HRTB. Biofilm detachment (erosion and sloughing) is a complex process affected by hydrodynamic conditions together with morphological and physiological characteristics of the biofilm [8, 42]. Suspended biomass changes were also observed at inflow rates (1.0–2.0 L h^{-1}) for all bioreactor rotation speed (5–30 min^{-1}) as a consequence of biofilm erosion (continuous release of smaller biofilm parts) [43].

The suspended biomass biosorption capacity ($q_{X.L}$) during heavy metal removal is presented in **Table 3**. The inflow rate had a more pronounced effect on the biosorption capacity than the bioreactor rotation speed. Nevertheless, highest bioreactor rotation speed (30 min^{-1}) decreased thickness of stagnant liquid layer at the biomass surface and provided facilitate condition for metal ion adsorption. The increase of the inflow rate to 1.0 and 2.0 L h^{-1} was related to the increase of biomass biosorption capacity. Microbial biomass concentration and content have a significant effect on the biosorption capacity. Therefore, higher biomass biosorption capacity was observed for inflow rates 1.0 and 2.0 L h^{-1} where higher microbial biomass concentration and biofilm erosion were observed (**Table 2**). Biofilm structure and extracellular polysaccharide content increase possibility for metal ion accumulation. Molecule of extracellular polysaccharide has high molecular mass and enhanced capability for metal ion bonding [13, 42, 44–47]. Due to the biofilm detachment observed for F = 2.0 L h^{-1} and n = 30 min^{-1} and release of microbial biomass with high amount of biofilm, biosorption capacity reached highest value of 83.27 mg g^{-1}, respectively.

Biological and hydrodynamic factors (content of extracellular polymers and cell physiological and morphological state of same microbial species) have influence on the suspended biomass

F (L h^{-1})	n (min^{-1})		
	5	15	30
0.5	1.08	0.95	1.02
1.0	1.59	1.74	2.48
2.0	2.67	2.89	5.11

Table 2. The suspended biomass concentration ($c_{x.L}$) changes at different combinations of bioreactor process parameters (n and F) during heavy metal removal process.

F (L h^{-1})	n (min^{-1})		
	5	15	30
0.5	28.45	18.51	33.75
1.0	33.59	48.79	71.62
2.0	58.73	58.01	83.27

Table 3. Metal ion sorption capacity ($q_{x.L}$) changes at different combinations of bioreactor process parameters (n and F) during heavy metal removal process.

biosorption capacity. Situation is more complex in mixed culture where different microbiological content and cell distribution also influence biosorption capacity. In addition, hydrodynamic conditions have also influence on all previous denominate biological factors [8]. Therefore, on the basis of these results, it is clear that biological hydrodynamic conditions in HRTB have a significant effect on the suspended biomass concentration and biosorption capacity (**Tables 2** and **3**).

3.3. Biofilm volumetric density and thickness along HRTB

Since the sampling point at 75% of reactor length was also used for introducing the temperature sensor, the biofilm thickness could be measured only at four sampling sites. The differences in biofilm thickness given by changing medium inflow rate (F = 0.5–2.0 L h^{-1}) and bioreactor rotation speed (n = 5–30 min^{-1}) are presented in **Table 4**. The biofilm thickness was in the range of 0.23–1.43 mm that is thinner than the literature data for mixed culture biofilm but thicker than monomicrobial culture biofilm thickness measured in previous research [25].

The biofilm thickness in the bioreactor L_f was mainly stabile for inflow rates 0.5 and 1 L h^{-1}, and only smaller biofilm parts were observed in the liquid phase as a consequence of the biofilm erosion process. This tendency was maintained until the inflow rate became 2 L h^{-1}. Afterward, hydrodynamic conditions and high metal load inhibited biofilm growth and decreased biofilm thickness. The resultant accumulation of metal ions had an impact on the biofilm, its strength, and its density. In these conditions intensive detachment of the biofilm was observed. The increase of the inflow rate produces thinner biofilm with higher density. Therefore, the outer biofilm layers are more sensitive to the shear stress and abrasion than the

F (L h^{-1})	n (min^{-1})	L_f (mm)		
		(0% L_{HRTB})	(50% L_{HRTB})	(100% L_{HRTB})
0.5	5	0.75	1.08	0.89
	15	0.89	0.73	0.81
	30	0.93	1.29	0.85
1.0	5	0.85	1.34	0.95
	15	0.92	1.43	0.84
	30	0.86	1.21	0.91
2.0	5	0.23	0.37	0.28
	15	0.35	0.28	0.25
	30	0.38	0.37	0.35

Table 4. Biofilm thickness changes (L_f) along HRTB at different medium inflow rates (F = 0.5–2.0 L h^{-1}) and bioreactor rotation speed (n = 5–30 min^{-1}) during heavy metal removal bioprocess.

inner biofilm layers. Moreover, outer biofilm layers can be released even at relatively small shear stress. After this, the detachment rate is considerably reduced [12, 47]. Thinner biofilms are less sensitive to process condition changes, which has positive influence on the process stability [44]. The impact of the biofilm detachment on the bioprocess was less pronounced from bioreactor inflow rate (**Table 4**).

Biofilm volumetric density ($c_{x.f}$) for F = 2.0 L h^{-1} and n = 30 min^{-1} was measured at the inlet and the outlet of the HRTB. The HRTB is characterized by concentration gradient along bioreactor, so consequently higher volumetric biofilm density was observed at the inlet of HRTB (59.7 ± 5.2 g L^{-1}) than at the outlet of HRTB (39.3 ± 4.4 g L^{-1}). Similar results were observed during previous investigation of metal ion removal in HRTB [38].

The reason for this finding might be that the substrate concentrations for microorganism growth decrease with bioreactor length. Higher volumetric biofilm density was related to increase the biofilm sorption capacity. Both properties are influenced by structure and content of biofilm. Differences in extracellular polysaccharide content affect the gradient of the linkage strength between cell clusters inside the biofilm. While cells on the surface of the biofilm grow relatively fast and do not accumulate, cells inside the biofilm have lower growth rates and produce more extracellular polysaccharides [13, 42, 44–46]. The extracellular polysaccharides affect the microbial sorption capacity by their content and molecular size. The outer biofilm layer exhibits higher porosity, resulting in easier metal ion access to deeper layers. Additionally, high-volumetric-density biofilms have higher sorption capacity than the low-density biofilms that are characterized by the low content of extracellular polysaccharides [48].

3.4. Biofilm application in removal of Co(II), Cr(IV), and Mn(II) from wastewater

After biofilm formation and characterization, investigation of biofilm sorption abilities in removal of Co(II), Cr(IV), and Mn(II) was done at different combinations of medium inflow rates and constant HRTB rotation speed. Results are presented as equilibrium metal ion concentration along HRTB in the liquid phase. Equilibrium metal ion concentration was reached after five residence time changes.

The metal ion concentrations along HRTB at different medium inflow rates (F = 0.5–2.0 L h^{-1}) and constant bioreactor rotation speed (n = 15 min^{-1}) are presented in **Figure 5** (Co(II) concentration **Figure 5A**, Cr(VI) concentration **Figure 5B**, Mn(II) concentration **Figure 5C**). Points represent measured values, while simulated values are represented with curves. Metal ion concentration changes along HRTB were simulated using one-dimensional diffusion-biosorption model and optimized parameter values [38]. The inflow of all metal ion (Co(II), Cr(VI), and Mn(II)) concentration was 0.125 g L^{-1}, respectively. Lower metal ion concentrations were detected at a first measuring point in the bioreactor (located at the place of medium inflow, 0% L_{HRTB}) because of medium dilution at this location in the HRTB.

Generally, increase in the inflow rate (*F*) caused increase of metal ion concentration along bioreactor. Higher inflow rate increased metal ion load in HRTB and concentration of metal ions in liquid phase. Metal ion concentration in biomass was in a dynamic equilibrium with metal ion concentration in the liquid phase. Biomass (solid phase) in bioreactor becomes saturated with metal ions and reaches maximum removal capacity. Consequence of biomass saturation is the decrease of metal ion concentration in the liquid phase (**Figure 5**).

As shown in previously performed hydrodynamic experiments in HRTB, medium flow in the bioreactor can be determined by plug-flow conditions [21]. These are attributed to the formation of temperature and/or concentration gradients along the reactor length [16]. Decrease in the metal ion concentration gradient along the bioreactor length in the second part of the HRTB (measurements points on 50% and 100% L_{HRTB}) confirmed assumption of the plug-flow condition in HRTB (**Figure 5**). The highest metal ion concentration measured near the place of medium inflow (measurement points 0% and 25% L_{HRTB}) inhibited biomass

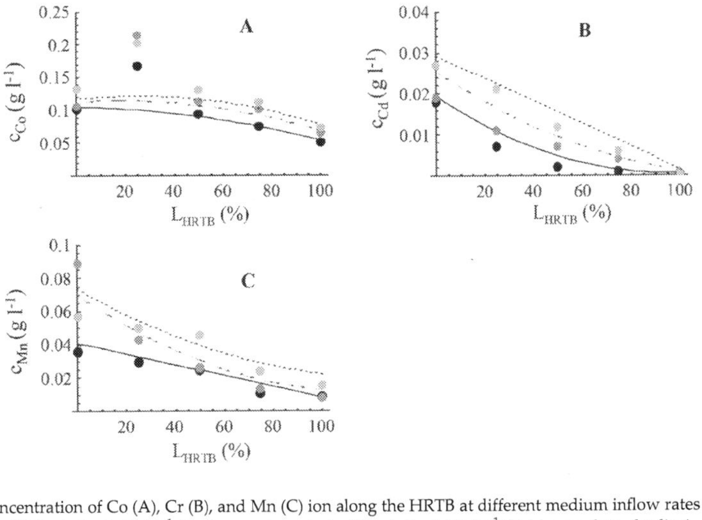

Figure 5. Concentration of Co (A), Cr (B), and Mn (C) ion along the HRTB at different medium inflow rates F = 0.5 L h^{-1} (black dots, solid line), F = 1.0 L h^{-1} (dark gray dots, dashed line), F = 2.0 L h^{-1} (light gray dots, dot line), and constant bioreactor rotation speed (n = 15 min^{-1}).

activity and produced a considerable deviations from plug-flow conditions. As was previously mentioned (in the Section 3.2), the biofilm biosorption is a complex process that is affected by hydrodynamic conditions as well as morphological and physiological characteristics of the biofilm [49, 50].

4. Conclusion

Microbial strains were isolated from heavy metal-contaminated surface sediments and selected due to their ability to grow in the presence of metal ions. The results obtained in this study proved technical feasibility of isolated strains to form biofilm in HRTB and to remove metal ions from contaminated water with concentrations up to 500 mg L^{-1}. The microbial removal ability was higher at lowest medium inflow rates of 0.5. When the inflow rate was in the range of 1.0–2.0 L h^{-1}, microbial removal ability was reduced.

Generally, the medium inflow rate had more pronounced effect on the bioprocess dynamics than bioreactor rotation speed. The biofilm biosorption capacity was reduced with decreased biofilm density. Similar trend shows suspended biomass biosorption capacity and suspended biomass concentration. The obtained results prove that HRTB can be successfully used for conducting the removal of heavy metals with isolated microbial strains.

Acknowledgements

This chapter is supported by Croatian Science Foundation project UIP-2014-09-1534 and SPECH-LRM-9158 for which the authors are grateful. Any findings, conclusions, and remarks are of authors only and do not necessarily reflect the standings of Croatian Science Foundation.

Author details

Tonči Rezić[1]*, Iva Rezić[2], Michaela Zeiner[3] and Božidar Šantek[1]

*Address all correspondence to: trezic@pbf.hr

1 Department of Biochemical Engineering, Faculty of Food Technology and Biotechnology, University of Zagreb, Zagreb, Croatia

2 Department of Applied Chemistry, Faculty of Textile Technology, University of Zagreb, Zagreb, Croatia

3 Department of Chemistry, University of Natural Resources and Life Sciences, BOKU – Vienna, Wien, Austria

References

[1] Deng L, Su Y, Su H, Wang X, Zhu X. Biosorption of copper (II) and lead (II) from aqueous solutions by nonliving green algae *Cladophora fasciculares*: equilibrium, kinetics and environmental effects. Adsorption. 2006;**12**:267–277.

[2] Kadirvelu K, Palanivel M, Kalpana R, Rajeswari S. Activated carbon from agricultural product, for the treatment of dyeing industry wastewater. Bioresour Technol. 2000;**74**:263–265.

[3] Pandey G, Jain RK. Bacterial chemotaxis toward environmental pollutants: role in bioremediation. Appl Environ Microbiol. 2002;**68**:5789–5795.

[4] Valko M, Morris H, Cronin MTD. Metals, toxicity and oxidative stress. Curr Med Chem. 2005;**12**:1161–1208.

[5] Gardea-Torresday JL, Tang L, Salvador JM. Copper adsorption by esterified and unesterified fractions of sphagnum peat moss and its humic substances. J Hazard Matter. 1996;**48**:191–206.

[6] Wang J, Chen C. Biosorbents for heavy metals removal and their future. Biotechnol Adv. 2009;**27**:195–226.

[7] Patterson JW. Industrial wastes reduction. Environ Sci Technol. 1989;**23**:1032–1038.

[8] Gavrilescu M. Removal of heavy metals from the environment by biosorption. Eng Life Sci. 2004;**3**:219–232.

[9] Gadd GM. Biosorption: critical review of scientific rationale, environmental importance and significance for pollution treatment. J Chem Technol Biotechnol. 2008;**84**:13–28.

[10] Singh R, Paul D, Jain RK. Biofilms: implications in bioremediation. Trends Microbiol. 2006;**14**:387–397.

[11] Kratochvil D, Volesky B. Advances in biosorption of heavy metals. Trends Biotechnol. 1998;**16**:291–300.

[12] Gavril M, Macoveanu M. Attached-growth process engineering in wastewater treatment. Bioproc Eng. 2000;**23**:95–106.

[13] Nicolella C, Loosdrecht MCM, Heijnen JJ. Wastewater treatment with particulate biofilm reactor. J Biotechnol. 2000;**80**:1–33.

[14] Gupta AB, Gupta SK. Simultaneous carbon and nitrogen removal from high strength domestic wastewater in an aerobic RBC biofilm. Wat Res. 2001;**35**:1714–1722.

[15] Moser A. Imperfectly mixed bioreactor systems. In: Moo-Young M (ed) Comprehensive biotechnology, vol. 2, Pergamon Press, Oxford; 1985. pp. 77–98.

[16] Moser A. Bioprocess technology kinetics and reactors. Springer, Wien New York; 1988.

[17] Kargi F, Eker S. Comparison of the performance of rotating perforated tubes and rotating biodisc biofilm reactors for wastewater treatment. Proc Biochem. 2001;**37**:1201–1206.

[18] Šantek B, Horvat P, Novak S, Mayr B, Moser A, Marić V. Mathematical modeling of mixing in horizontal rotating tubular bioreactor: "simple flow" model. Bioproc Eng. 1996;**14**:195–204.

[19] Šantek B, Horvat P, Novak S, Mayr B, Moser A, Marić V. Mathematical modeling of mixing in horizontal rotating tubular bioreactor: "simple flow" model. Bioproc Eng. 1996;**14**:223–229.

[20] Šantek B, Horvat P, Novak S, Moser A, Marić V. Studies on mixing in a horizontal rotating tubular bioreactor. Part I: optimisation of adjustable parameters for "spiral flow" model. Bioproc Eng. 1998;**18**:467–473.

[21] Šantek B, Horvat P, Novak S, Moser A, Marić V. Studies on mixing in a horizontal rotating tubular bioreactor. Part II: prediction systems for adjustable parameters of "spiral flow" model. Bioproc Eng. 1998;**19**:19–28.

[22] Šantek B, Horvat P, Novak S, Moser A, Marić V. Studies on mixing in a horizontal rotating tubular bioreactor. Part III: influence of liquid level and distance between the partition walls on prediction systems for adjustable model parameters. Bioproc Eng. 1998;**19**:91–102.

[23] Ivančić M, Šantek B, Novak S, Marić V. Fermentative bioconversion in a horizontal rotating tubular bioreactor. Proc Biochem. 2004;**39**:995–1000.

[24] Slavica A, Šantek B, Novak S, Marić V. Microbial acetate oxidation in horizontal rotating tubular bioreactor. J Biosci. 2004;**29**:169–177.

[25] Rezić T, Šantek B, Novak S, Marić V. Heterotrophic cultivation of *Paracoccus denitrificans* in a horizontal rotating tubular bioreactor. World J Microbiol Biotechnol. 2007;**23**:987–996.

[26] Rezić T, Šantek B, Novak S, Marić V. Comparison between the heterotrophic cultivation of *Paracoccus denitrificans* in continuous stirred tank reactor and horizontal rotating tubular bioreactor. Proc Biochem. 2006;**41**:2024–2028.

[27] Ivančić M, Šantek B, Novak S, Horvat P, Marić V. Bioprocess kinetics in a horizontal rotating tubular bioreactor. Bioprocess Biosyst Eng. 2004;**26**:169–175.

[28] Fries J, Getrost H. Organic reagents for trace analysis. Merck Laboratory Press, Darmstadt; 1977.

[29] Venkataraman R, Ramanujam TK. A study on microbiology of biological film layer in rotating biological contactors. Bioproc Eng. 1998;**18**:181–186.

[30] Wilke CR, Chang P. Correlation of diffusion coefficients in dilute solutions. AIChE J. 1955;**1**:264–270.

[31] Horn H, Morgenroth E. Transport of oxygen, sodium chloride and sodium nitrate in biofilms. Chem Eng Sci. 2006;**61**:1347–1356.

[32] Stewart PS. Diffusion in biofilms. J Bacteriol. 2003;**185**:1485–1491.

[33] Garrido JM, van Benthem W, van Loosdrecht MCM, Heijnen JJ. Influence of dissolved oxygen concentration on nitrite accumulation in a biofilm airlift suspension reactor. Biotechnol Bioeng. 1997;**53**:168–178.

[34] Freundlich H. Über die Adsorption in Lösungen. [On the adsorption in solutions; article in German] Z Phys Chem. 1907;**57**:385–470.

[35] Arora S, Dhaliwal SS, Kukreja VK. Application of orthogonal collocation on finite elements for solving non-linear boundary value problems. Appl Math Comput. 2006;**18**:516–523.

[36] Villadsen J, Stewart WE. Solution of boundary value problems by orthogonal collocation. Chem Eng Sci. 1967;**22**:1483–1501.

[37] Leitão A, Rodrigues A. Modeling of biodegradation/adsorption combined processes in fixed-bed biofilm reactors: effects of the intraparticle convective flow. Chem Ing Sci. 1996;**51**:4595–4604.

[38] Rezic T, Zeiner M, Santek B, Novak S. Mathematical modeling of Fe(II), Cu(II), Ni(II) and Zn(II) removal in a horizontal rotating tubular bioreactor. Bioprocess Biosyst Eng. 2011;**34**:1067–1080.

[39] Mikac N, Foucher D, Kwokal Ž, Barišiæ D. Mercury and radionuclides in sediments of the Kaštela bay (Croatia)– evaluation of the sediment pollution history. Croat Chem Acta. 2006;**79**:85–93.

[40] Bogner D. Surface sediments and influence of the physico-chemical characteristics of the sediment on the distribution of heavy metals in the Kaštela bay, in Croatian [MSc thesis]. University of Zagreb, Zagreb; 1996.

[41] Stegnar P, Vukadin I, Smodiš B, Vakselj A, Prosenc A. Trace elements in sediments and organisms from Kaštela bay. J. Etud. Pollut. CIESM. 1980;**5**:596–600.

[42] Peyton BM. Effects of shear stress and substrate loading rate on *Pseudomonas aeruginosa* biofilm thickness and density. Wat Res. 1996;**30**:29–36.

[43] Brading MG, Jass J, Lappin-Scott HM. Dynamics of bacterial biofilm formation. In: Lappin-Scott HM, Costerton JW (eds.) Microbia biofilms, Cambridge University Press, Cambridge; 1995. pp. 46–63.

[44] Liu Y, Tay J. The essential role of hydrodynamic shear force in the formation of biofilm and granular sludge. Wat Res. 2002;**36**:1653–1665.

[45] Stoodley P, Sauer K, Davies DG, Costerton JW. Biofilms as complex differentiated communities. Annu Rev Microbiol. 2002;**56**:187–209.

[46] Stoodley P, Hall-Stoodley L, Lappin-Scott HM. Detachment, surface migration and other dynamic behavior in bacterial biofilms revealed by digital time-lapse imaging. Methods Enzymol. 2000;**337**:306–319.

[47] Characklis WG. Biofilm processes. In: Characklis WG, Marshall KC (eds.) Biofilms, Wiley, New York; 1990. pp. 93–130.

[48] Cloirec P, Andres Y, Faur-Brasquet C, Gerente C. Engineered biofilms for metal ion removal. Rev Environ Sci Biotechnol. 2003;**2**:177–192.

[49] Quintelas C, Fonseca B, Silva B, Figueiredo H, Tavares T. Treatment of chromium(VI) solutions in a pilot-scale bioreactor through a biofilm of *Arthrobacter viscosus* supported on GAC. Bioresour Technol. 2009;**100**:220–226.

[50] Van Hullebusch ED, Zandvoort MH, Lens PNL. Metal immobilization by biofilms: mechanisms and analytical tools. Rev Environ Sci Bio Technol. 2003;**2**:9–33.

Chapter 10

Electrocoagulative and Biological Treatment of Laundry Wastewater

Terelle Ramcharan and Ajay Bissessur

Additional information is available at the end of the chapter

http://dx.doi.org/10.5772/67525

Abstract

The greater demand for potable water, both locally and worldwide, has directed a huge interest amongst researchers to investigate the possibility of recycling and reusing wastewater from laundry run-offs. The advantage of using recycling wastewater from such sources is mainly due to the fact that these bulk volumes of wastewater are considered to be less chemically polluted in comparison to those discarded from industrial effluents and wastewater sources. Almost all laundry detergents contain surfactants, whose main function serves to remove dirt/soil from contaminated items. Thus, an analysis of the surfactant levels before and after a treatment process is important to confirm that the surfactant has in fact carried out its intended purpose. Electrocoagulative treatment of wastewater, a well-researched and well-documented clean-up process that involves the production of aluminium hydroxy species by oxidation of aluminium metal upon the application of a controlled voltage which adsorbs fine particulate matter and pollutants from the wastewater has been investigated as a clean-up application to the treatment of laundry wastewater. The use of a biological treatment process which entails treating the wastewater with aerobic bacterial specie specifically designed to degrade fats, lipids, protein, detergents and hydrocarbons has also been investigated.

Keywords: biological, biospinners, electrocoagulation, laundry wastewater, linear alkylbenzene sulfonates

1. Introduction

The composition of laundry detergents is generally complex due to the numerous factors that have to be taken into consideration to ensure fresh clean garments at the end of the wash process. Sodium dodecylbenzene sulfonate, more commonly known as SDS or linear alkylbenzene sulfonates (LAS), is the most abundant anionic surfactant utilised in laundry detergents due to its excellent performance in removing water insoluble substances such as greasy and oily stains. As a commercial commodity, LAS is sold as a sodium salt which contains a mixture of homologues that has between 10 and 14 linear carbon atoms with a phenyl group attached to the linear alkyl chain and the sulfonate anion as shown in **Figure 1** [1–6].

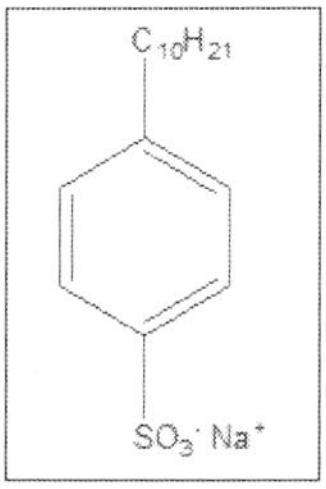

Figure 1. Chemical structure of sodium dodecylbenzene sulfonate (SDS).

The rapid biodegradation of LAS compounds especially under aerobic conditions consumes a large amount of bio-available oxygen that significantly increases the chemical oxygen demand, thus negative impacting on the environment and organisms persisting within that system [4, 5]. Oxidation of LAS by oxygen results in the formation of sulfophenylcarboxylic acid (SPC) that comprises one of the main products of biodegradation [7–9].

2. Quantification of LAS by Ultraviolet-Visible spectrophotometry

Ultraviolet-Visible (UV-Vis) spectrophotometry is one of the commonly used techniques for the quantification of surfactants, whereby the method of determination of anionic surfactants entails the use of a cationic dye that complexes with the anionic surfactant through the mechanism of ion association as shown in **Figure 2** [10, 11].

Valuable structural information by mass spectrometric (MS) detection often allows for the qualitative analysis of surfactants [12]. Analysis of ethoxylated surfactants using soft ionisation techniques such as electrospray ionisation (ESI) and atmospheric pressure chemical ionisation (APCI) determines analytes in a cationized molecular state [13]. The use of mass spectrometry and addition of a volatile reagent like ammonium salt, for example, ammonium acetate that suppresses the formation of alkali salts improves the accuracy of LAS determinations. The determination of non-ionic surfactants is possible via the application of positive or negative ionization modes for ESI and APCI, with the best response obtained using the positive ion mode [13].

Liquid chromatography-mass spectroscopy (LC-MS) is a powerful analytical technique, that is an applied qualitative detection method for non-ionic surfactants as reported by many researchers [12, 14–17]. In addition, a direct application of gas chromatography-mass spectroscopy (GC-MS) is used in the analysis of non-ionic surfactants; however, this method is limited in its

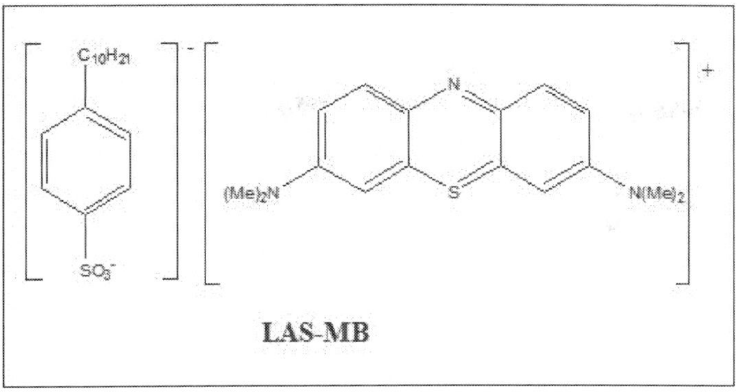

Figure 2. Ion association complex formed between LAS and methylene [10].

application due to the derivatization requirement for long ethoxy chain containing surfactants [17]. The use of solid-phase extraction (SPE) and GC-MS for the direct analysis of APEs is carried out, whereby a graphitized carbon black SPE cartridge and use of methanol/dichloromethane solvent system was implemented [18]. The use of ethyl violet and acridine orange dyes has been reported by researchers for extraction of anionic surfactants [19, 20]. Specifically, toluene and benzene solvents have been used for extraction of LAS complexes, which is deemed less toxic than chloroform, and have therefore been reported as a recommended replacement to the methylene blue method [19, 20]. High-performance liquid chromatography (HPLC) is a commonly applied technique for LAS determination and detection which includes ultraviolet (UV), fluorescence (FL), diode-array detection (DAD) and mass spectroscopy (MS).

Another method for the analysis of LAS by an ion-pair SPE technique and HPLC has been developed [21]. Extraction of LAS using C_8, C_{18} and multiwall carbon-nanotubes was investigated and samples were quantified by reversed-phase HPLC using a C_8 column and UV detection with isocratic elution at a retention time of 15 min using a methanol/water mobile phase containing 5 mM sodium acetate [21]. Quantification of LAS in environmental samples by HPLC-FL has been developed which entails Soxhlet extraction of the sample with gradient elution, retention time of 22 min and application of mobile phases, which include acetonitrile, water, triethylamine and acetic acid [22, 23]. Quantification of LAS in sewage sludge samples using HPLC-FL with a C_8 column with microwave-assisted extraction is used for sample preparation. A comparison of separation of LAS using HPLC-FL and HPLC-DAD showed no significant difference between the two sets of results and that usage of either a FL or DAD detector are applicable [24]. HPLC-MS is considered the most accurate method for determination of LAS as it permits for both a qualitative and quantitative analysis of LAS [3, 5]. GC-MS is less often used for analysis of LAS, as this method would require derivatization of LAS into a volatile compound [25]. Quantification of anionic surfactants and inorganic constituents' viz., phosphates, silicates and zeolite, has been analysed by Inductively coupled plasma-optical emission spectroscopy (ICP-OES) [26]. Specifically, alkylbenzene sulfonates and alkyl sulphates were determined due to their ability to precipitate upon addition of calcium ions [26]. Non-ionic surfactants that are used widely in domestic and industrial detergents [27] are represented by two major classes, which include alcohol ethoxylates (AE) and alkylphenol ethoxylates (APEOs) [28]. The most common non-ionic surfactants used in detergents are octylphenol ethoxylate (OPEO) and nonylphenol ethoxylate (NPEO) as shown in **Figure 3** [29, 30].

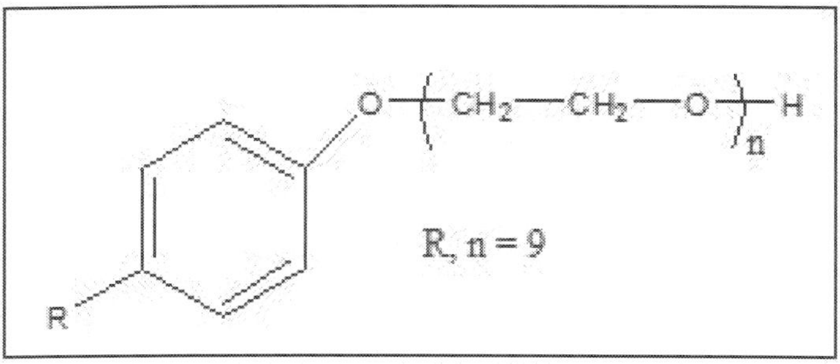

Figure 3. Structure of nonylphenol ethoxylate [31, 32].

The toxicity of the surfactant is dependent on the length of the ethoxy chain. A more toxic behaviour is known to be displayed by APEOs with a shorter ethoxy chain (typically <4) when compared to longer ethoxy chain length APEOs (typically >10) [18]. APEOs can be degraded under both anaerobic and aerobic conditions, thus leading to the biotransformation of APEO into lipophilic metabolites of APEO [33]. The most common degradation products of APEOs include nonylphenol, octylphenol and mono- and diethoxylated compounds of NPEO and OPEO [33, 34], which are deemed toxic and have been found to be persistent in the environment, thus causing endocrine disrupting effects amongst aquatic organisms [34–36]. Other contributing important ingredients found in laundry detergents include builders and anti-foaming agents. A common zeolite-based builder, sodium aluminium silicate, is often used as a builder in laundry detergents to reduce water hardness, while polydimethylsiloxane acts as an anti-foaming reagent.

3. Application of biological and electrocoagulative treatment methods to laundry wastewater

The separation of the solid matrices from the liquid matrices forms the basis for treatment of wastewater, which is most commonly achieved through coagulation-flotation methods [37]. During coagulative processes, an alteration of the surface properties of the individual particles occurs and this permits transformation into larger particles [38]. Inorganic salts of aluminium, iron or calcium are commonly used in coagulation processes [39]. In the coagulation process, small particles may form which decrease the efficiency in removal of pollutants from the wastewater streams and for this reason, flocculent agents are commonly used in conjunction with coagulation agents [40]. The efficiency of coagulation is enhanced by an increase in flocculation through accumulation of particles into larger settleable masses [38]. Polymer-based flocculants are commonly used for this purpose as a result of their large surface area,

http://dx.doi.org/10.5772/67525

hence enabling the particles to group and settle, thus facilitating easy removal of pollutants from the wastewater.

Biological treatments have been mainly applied to the treatment of industrial effluent wastewater. The advantages associated with biological treatment of wastewater include a decreased amount of toxic and harmful chemicals coupled with an easy to implement green process [41]. Waste from effluents is recycled into an organism-based biomass through biological treatment, and can be easily disposed of naturally into the environment [41]. Major disadvantages associated with biological treatment of wastewater include:

a. large space requirement for the storage of biological waste,

b. longer time periods required for effluent treatment in comparison to chemical treatment,

c. limitation in its application to a wide range of effluents [41].

In the application of biological treatment of wastewater, addition of a specific strain of bacteria to the wastewater is the main thrust of the system that subsequently targets specific oxidation and degradation of pollutants.

Biological wastewater treatment is often seen as an environmental friendly method, as there are generally no added chemicals involved. Some of the major concerns with regard to biological treatment of wastewater include the longer time periods for treatment, a larger surface area required and the addition of specialised bacteria for the specific degradation of pollutants. Chan demonstrated a method for treatment of laundry effluent through a combination of biological and chemical treatment methods [41].

The laundry effluent was treated biologically prior chemical treatment. This treatment method permitted the production of high-quality water that could be used for activities such as flushing and cleaning which reduced the consumption of water by the launderette. The quality of the water was assessed by measuring the following parameters: pH, DO, SS, COD and total surfactant concentration.

Electrocoagulation is often implemented as the primary treatment for wastewater due to its efficient pollutant removal as well as its safe and environmental friendly nature. Electrocoagulation involves the dissolution of sacrificial anodes due to the application of electric current. Aluminium and iron are the most generic anodes used for this purpose.

$$Al(s) \rightarrow Al^{3+}(aq) + 3e^- \quad (1)$$

$$3H_2O\ (\ell) + 3e^- \rightarrow H_2(g) + 3OH^-(aq) \quad (2)$$

Eqs. (1) and (2) represent the reactions taking place at the anode and cathode, respectively. The resultant metal ion reacts with hydroxide in the wastewater to form various metal hydroxides.

$$Al^{3+}(aq) + 3H_2O\ (\ell)\ (aq) \rightarrow Al(OH)_3 + 3H^+(aq) \quad (3)$$

$$Al(OH)_3(aq) + OH^-(aq) \rightarrow Al(OH)_4{}^-(aq) \quad (4)$$

Eqs. (3) and (4) represent the generation of aluminium hydroxy species during electrocoagulation.

Treatment of wastewater by electrocoagulation is known to effectively remove heavy metals, minerals and dyes from wastewater streams, hence making it a good treatment method for laundry wastewater. A high removal efficiency of organic compounds is obtained due to the various mechanisms that occur in the electrocoagulation cell. The pollutants adsorb onto the different aluminium hydroxy species depending on the chemical structure of the pollutant.

The hydrogen gas produced at the cathode induces flotation of the hydroxy species, hence allowing for a quick and efficient removal of pollutants from the wastewater. Aside from the production of aluminium hydroxy species, other mechanisms in the electrocoagulation cell occur which increases the efficient removal of pollutants from the wastewater stream. Reactions at the surface of the cathode also remove carbonate salts, which is abundant in laundry wastewater.

$$HCO^{3-}(aq) + OH^-(aq) \rightarrow CO_3{}^{2-}(aq) + H_2O\ (\ell) \tag{5}$$

$$Ca^{2+}(aq) + CO_3{}^{2-}(aq) \rightarrow CaCO_3(s) \tag{6}$$

$$Mg^{2+}(aq) + CO_3{}^{2-}(aq) \rightarrow MgCO_3(s) \tag{7}$$

Eqs. (5)–(7) represent the removal of carbonate from the wastewater as salts of calcium and magnesium. Laundry wastewater is also known to contain chloride salts. Electrolysis generates molecular chlorine, which can lead to the formation of hypochlorous acid and hypochlorite ions as shown in Eqs. (8)–(10). These species contain a relatively high oxidative potential, which allows for further degradation of organic pollutants in the wastewater stream.

$$2Cl^-(aq) \rightarrow Cl_2(g) + 2e^- \tag{8}$$

$$Cl_2(g) + H_2O \rightarrow HOCl\ (aq) + H^+(aq) + Cl^-(aq) \tag{9}$$

$$HClO(aq) \rightarrow ClO^-(aq) + H^+(aq) \tag{10}$$

In research presented by many scientists, electrocoagulation is described as the treatment of laundry effluent [42–44]. Iron and aluminium electrodes are used for electrocoagulation; however, aluminium electrodes had a greater efficiency in removal of pollutants from the laundry wastewater. Some investigations applied an ultrasonic bath during electrocoagulation which had a profound effect on the efficiency of the removal [42].

Over time, the formation of an inhibiting film due to high voltages applied to the electrodes impacts negatively on the efficiency of electrocoagulation. The measured parameters of phosphorous levels, detergent, COD, turbidity and conductivity in the laundry wastewater before and after the process of electrocoagulation are good indicators of the effectiveness of the electrocoagulative process [42–44].

Electrocoagulation using aluminium electrodes, as shown in **Figure 4**, has been applied as a method for treatment of wastewater obtained from a textile industry aimed at the removal of dye substances from wastewater [45, 46]. This method has accounted for a 99% efficiency in removal of the dye substances, measured by determination of the COD before and after treatment [45, 46]. The removal of heavy metals such as nickel, copper, zinc and chromium from synthetic and industrial wastewater by electrocoagulation using aluminium electrodes

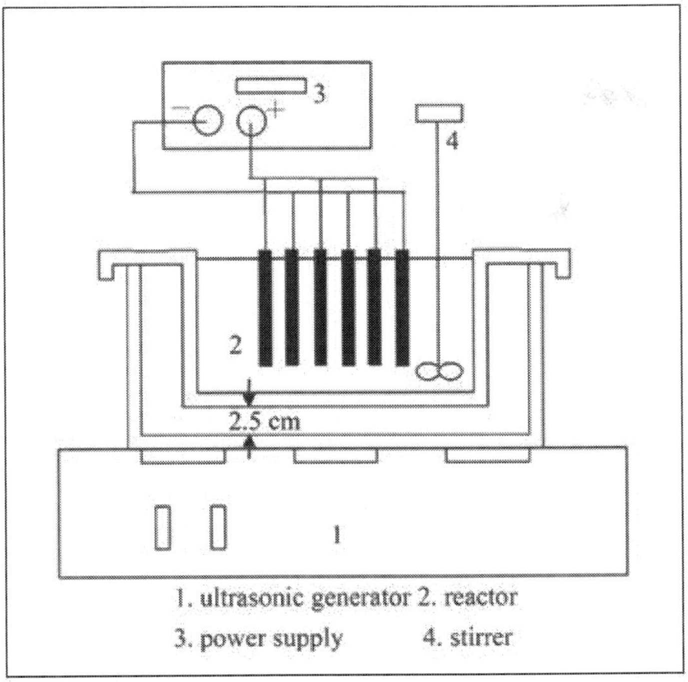

Figure 4. Illustration of an electrocoagulation cell adapted from Wang et al. [42].

has been widely applied. An added advantage of electrocoagulation in addition to removal of heavy metals from wastewater stream also significantly decreased the COD [47]. In a research study by Ramcharan and Bissessur, a comparison of electrocoagulation and biological treatment of Laundry Wastewater (LWW) was reported [48]. The surfactant concentration, chemical oxygen demand and total dissolved solids were the general water guideline parameters used to assess the success of the treatment system. The wastewater was characterised after each wash and rinse cycle discharged from a domestic washing machine and are referred to as first wash cycle wastewater (W1), first rinse cycle wastewater (R1) and second rinse cycle wastewater (R2). The two major parameters, which influenced the above treatment methods, were the period allocated for treatment and the suitability of each treatment method to a variety of wastewater matrixes. The successful treatment of R1 and R2 was obtained using the biological method, while electrocoagulation was successful for W1, R1 and R2 (**Figure 5**). The sample matrix of W1 was not compatible for biological treatment, as the bacterium was not able to cultivate under such harsh conditions. Aeration of W1 proved to decrease the concentration of the surfactant because SDS is susceptible to degradation under oxidative conditions.

Degradation of the bacteria is imminent upon exposure to the strongly basic pH of the first wash laundry wastewater, which increased the organic content thereby increasing the COD in laundry wastewater from the first wash during biological treatment.

The dominance of the electrocoagulative treatment method over the biological method of LWW is further supported by the COD levels attained as shown in **Figure 5**. It is clearly evident that upon treatment of W1, a gradual increase in the COD levels occurs over a prolonged period of time. The highly alkaline nature of the wastewater induces breakdown of bacterial cells, thus implementing an increase in the organic content and thereby consequently

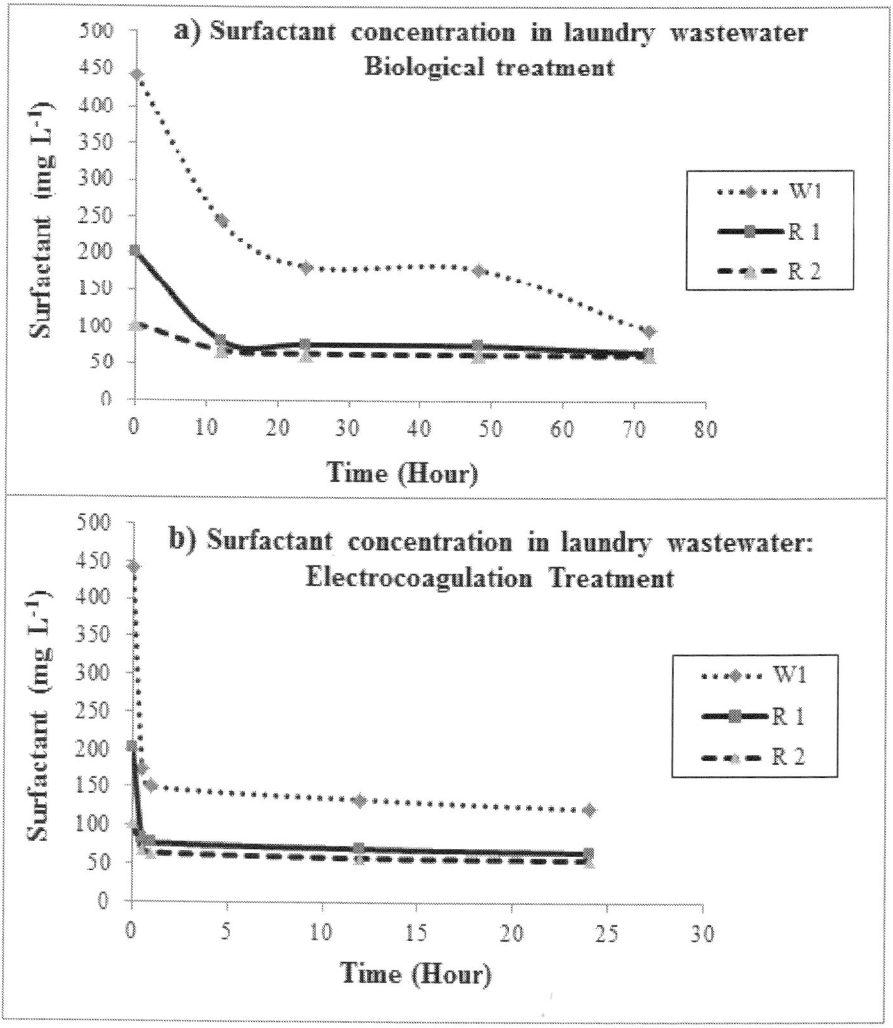

Figure 5. Decrease in surfactant concentration after application of (*a*) *biological treatment* and (b) *electrocoagulative treatments* to laundry wastewater from the first wash (W1), first rinse stage (R1) and second rinse stage (R2). Reproduced from Ref. [48].

causing an increase in the COD level of W1. However, a marked decrease in COD level occurred during the implementation of the electrocoagulative technique as shown in **Figure 5**. Finally, the persistence of LAS in solution is directly linked to the COD level. The effective removal of LAS by the electrocoagulative treatment caused a marked decrease in the organic content present; thus, a rapid decrease in the COD is observed especially for R2 in the initial onset (within the 5 minutes of implementation) of electrocoagulation as shown in **Figure 6**.

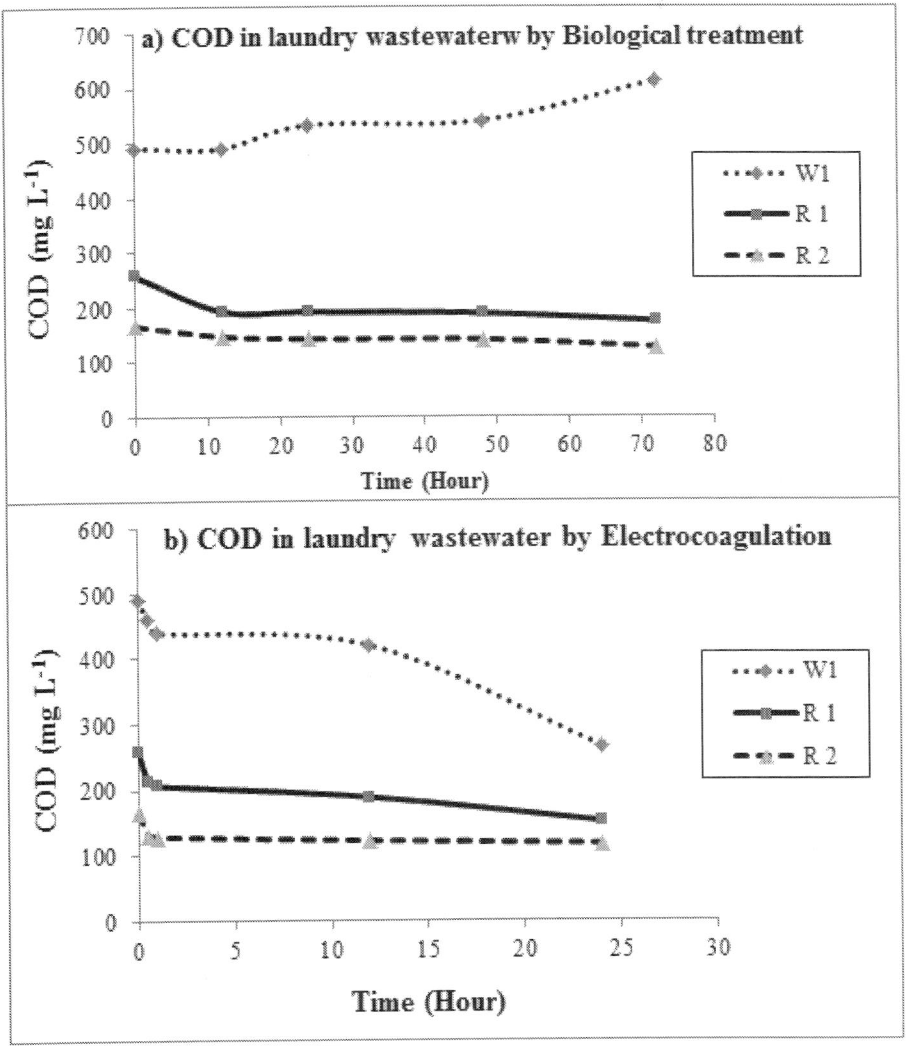

Figure 6. COD Levels of laundry wastewater samples after first wash (W1), first rinse (R1) and second rinse (R2) cycles when subjected to (a) *biological treatment* and (b) *electrocoagulation*. Reproduced from Ref. [48].

TDS levels at the different wash and rinse cycles of LWW showed an increasing trend when treated biologically, while the electrocoagulation method of treatment for LWW showed a decrease in the TDS levels as shown in **Figure 7**. This is chiefly due to the quick polymeric generation of aluminium hydroxide species during electrocoagulation had allowed for adsorption of SDS in LWW whilst promoting effective TDS removal through settlement of the polymeric floc generated.

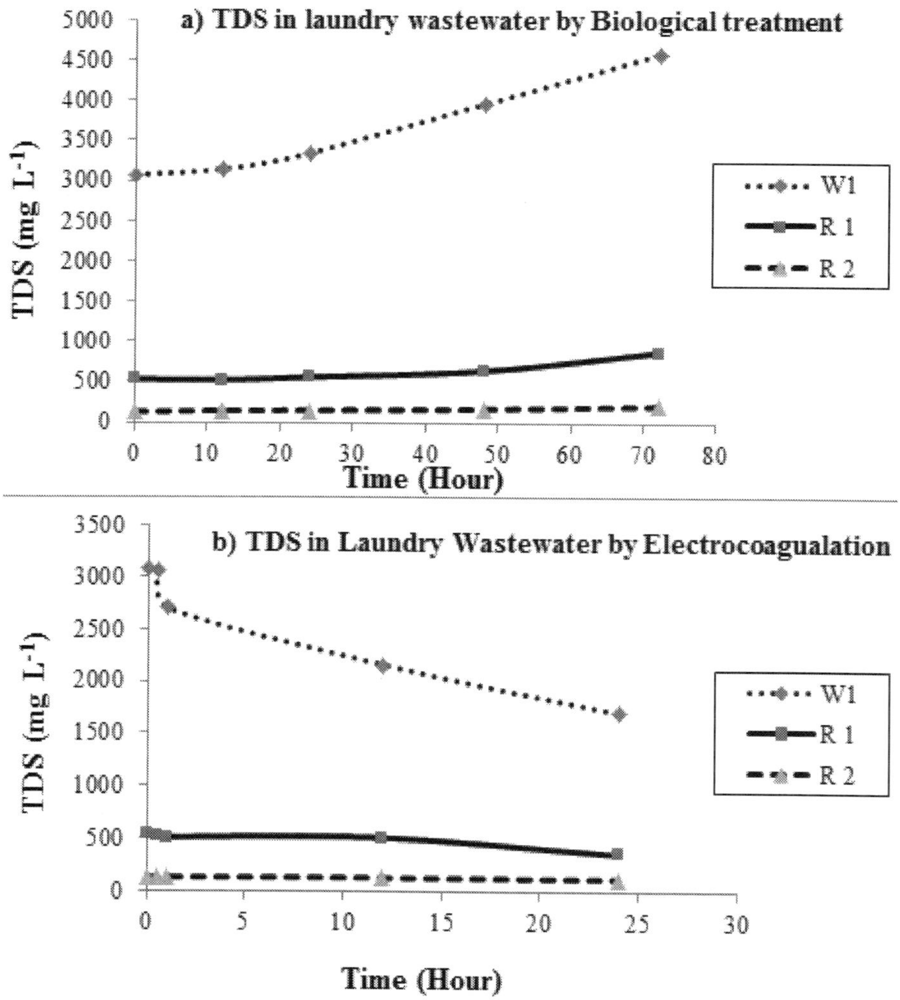

Figure 7. TDS levels in laundry wastewater for first wash (W1), first rinse (R1) and second rinse samples (R2) after (a) *biological treatment* and (b) *electrocoagulation*. Reproduced from Ref. [48].

Supporting kinetic data is pivotal when implementing pilot wastewater treatment systems. The adsorption kinetics is one of the important parameters used to assess sustainability of the treatment system. A kinetic study on the adsorption capacity of the aluminium hydroxy species was investigated by Ramcharan and Bissessur [48]. The Ho pseudo second-order expression was used to evaluate the adsorption capacity for surfactant removal in laundry wastewater from the first wash, first rinse cycle and second rinse cycle as shown in Eq. (11) below. A second-order reaction was observed from the plot of t/q_t vs. t shown in **Figure 8** with R^2 values >0.99.

$$\frac{t}{q_t} = \frac{1}{k_2 q_e^2} + \frac{t}{q_e} \tag{11}$$

The percentage of efficiency of adsorption (% E) was based on calculations using Eq. (13) below, where C_0 and C corresponds to the initial and specific concentration of the surfactant at time t. The values of the adsorption efficiency at equilibrium (q_e) and rate of adsorption (k_2) was based on calculations using Eqs. (13) and (14), respectively. The rate of adsorption of the surfactants is significantly lower for laundry wastewater discharged from the first wash as compared to laundry wastewater from the first and second rinses as shown in **Table 1**. It can be easily inferred that a reduced amount time is required for the treatment of laundry wastewater disposed after the first and second rinses.

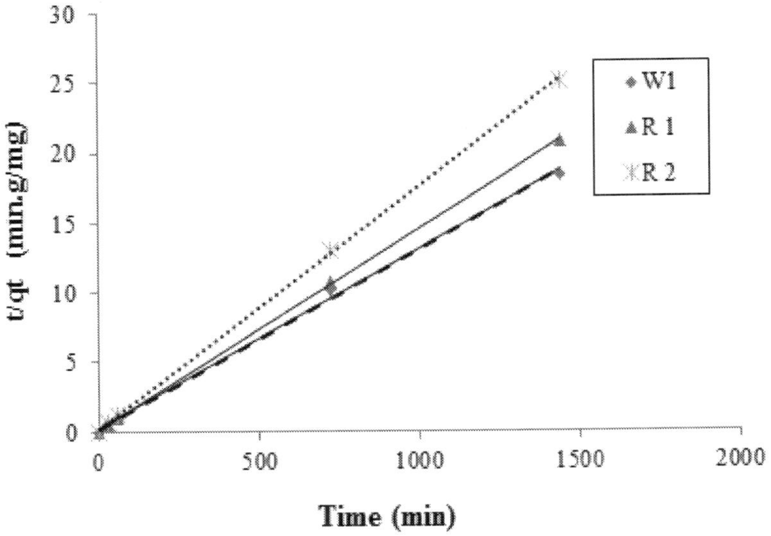

Figure 8. A Plot of t/q_t vs. t showing second-order reaction kinetics for the adsorption capacity of surfactant by aluminium hydroxy species. Reproduced from Ref. [48].

Lagergren parameter	W1	R1	R2
Experimental q_e	77.60	67.27	60.69
Calculated q_e	77.52	68.97	57.47
R^2	0.997	0.999	0.999
k_2	8.53×10^{-4}	2.53×10^{-3}	2.21×10^{-3}
Reproduced from Ref. [48].			

Table 1. The Lagergren parameters for adsorption of surfactants by aluminium hydroxy species.

$$\% E = \frac{C_0 - C}{C_0} \times 100 \tag{12}$$

$$q_e = \frac{1}{\text{Slope}} \tag{13}$$

$$k_2 = \frac{\text{Slope}^2}{\text{Intercept}} \tag{14}$$

4. Conclusions

The application of electrocoagulative and biological treatment methods effectively decreased the amount of surfactant concentration in laundry wastewater after all rinsing stages. In comparison, the electrocoagulative technique was found to be a more efficient treatment method of the two due to its ability to reduce the levels of the surfactant, COD and TDS over a considerably shorter period of time and its ability to be applied to a wider range of wastewater samples. A modification to the electrocoagulation treatment process whereby the addition of Biospinners® was carried out and was found to further reduce the levels of the surfactant, COD and TDS within the same applied period of time. Modification due to addition of Biospinners was shown to increase aeration and surface area, and facilitated the removal of an overlaying film of aluminium hydroxy species formed on the electrodes. The adsorption of LAS by aluminium hydroxy species was found to take place at a lower rate for W1, in comparison to R1 and R2 as shown by the kinetics in this study. From this, it is evident that there is a need for isolated treatments of laundry wastewater W1, R1 and R2, thus ensuring a reduced period of treatment and also ensuring the total output cost of the treatment method is kept to a minimum.

Author details

Terelle Ramcharan and Ajay Bissessur*

*Address all correspondence to: bissessura@ukzn.ac.za

School of Chemistry and Physics, University of KwaZulu-Natal, Durban, South Africa

References

[1] Santos J. L., Aparicio I., and Alonso E., *A new method for the routine analysis of LAS and PAH in sewage sludge by simultaneous sonication-assisted extraction prior to liquid chromatographic determination.* Analytica Chimica Acta, 2007. **605**: pp. 102–109.

[2] Schleheck D., Netzer F. v., Fleischmann T., Rentsch D., Huhn T., Cook A. M., and Kohler H. P. E., *The missing link in linear Alkylbenzenesulfonate surfactant degradation: 4-sulfoacetophenone as a transient intermediate in the degradation of 3-(4-Sulfophenyl)Butyrate by comamonas testosteroni KF-1.* Applied and Environmental Microbiology, 2010. **76**: pp. 196–202.

[3] Yangxin Y. U., Jin Z., and Bayly A. E., *Development of surfactants and builders in detergent formulations.* Chinese Journal of Chemical Engineering, 2008. **16** (4): pp. 517–527.

[4] Akyuz M., and Roberts D. J., *Determination of linear alkylbenzene sulphonates and their biodegradation intermediates by isocratic RP-HPLC.* Turkish Journal of Chemistry, 2006. **26**: pp. 669–679.

[5] Wangkarn S., Soisungnoen P., Rayanakorn M., and Grudpan G., *Determination of linear alkylbenzene sulfonates in water samples by liquid chromatography–UV detection and confirmation by liquid chromatography–mass spectrometry.* Talanta, 2005. **67**: pp. 686–695.

[6] Asok A. K., and Jisha M. S., *Biodegradation of the anionic surfactant linear alkylbenzene sulfonate (LAS) by autochthonous Pseudomonas sp.* Water, Air, & Soil Pollution, 2012. **223** (8): pp. 5039–5048.

[7] Sarrazin L., Arnoux A., Reboullion P., and Monod J. L., *Biodegradation of linear alkylbenzene-sulfonates (LAS) in briny water and identification of metabolites using HPLC analysis by direct injection of samples.* Toxkological and Environmental Chemis, 1997. **58**: pp. 209–216.

[8] Navas J. M., GonzÁlez-Mazo E., Wenzel A., GÓmez-Parra A., and Segner H., *Linear alkylbenzene sulfonates and intermediate products from their degradation are not estrogenic.* Marine Pollution Bulletin, 1999. **38**(10): pp. 880–884.

[9] González-Mazo E., and Gómez-Parra A., *Monitoring anionic surfactants (LAS) and their intermediate degradation products in the marine environment.* TrAC Trends in Analytical Chemistry, 1996. **15**(8): pp. 375–380.

[10] Jurado E., Fernandez-Serrano M., Nunez-Olea J., Luzon G., and Lechuga M., *Simplified spectrophotometric method using methylene blue for determining anionic surfactants: Applications to the study of primary biodegradation in aerobic screening tests.* Chemosphere, 2006. **65**: pp. 278–285.

[11] Chitikela S., Dentel S. K., and Allen H. E., *Modified method for the analysis of anionic surfactants as methylene blue active substances.* Analyst, 1995. **120**(7): pp. 2001–2004.

[12] Ferguson P. L., Iden C. R., and Brownawell B. J., *Analysis of alkylphenol ethoxylate metabolites in the aquatic environment using liquid chromatography-electrospray mass spectrometry.* Analytical Chemistry, 2000. **72**(18): pp. 4322–4330.

[13] Frömel T., and Knepper T. P., *Mass spectrometry as an indispensable tool for studies of biodegradation of surfactants*. TrAC Trends in Analytical Chemistry, 2008. **27**(11): pp. 1091–1106.

[14] Shao B., Hu J.-Y., and Yang M., *Determination of nonylphenol ethoxylates in the aquatic environment by normal phase liquid chromatography–electrospray mass spectrometry*. Journal of Chromatography A, 2002. **950**(1–2): pp. 167–174.

[15] Eadsforth C. V., Sherren A. J., Selby M. A., Toy R., Eckhoff W. S., McAvoy D. C., and Matthijs E., *Monitoring of environmental fingerprints of alcohol ethoxylates in Europe and Canada*. Ecotoxicology and Environmental Safety, 2006. **64**(1): pp. 14–29.

[16] Andreu V., Ferrer E., Rubio J. L., Font G., and Picó Y., *Quantitative determination of octylphenol, nonylphenol, alkylphenol ethoxylates and alcohol ethoxylates by pressurized liquid extraction and liquid chromatography–mass spectrometry in soils treated with sewage sludges*. Science of the Total Environment, 2007. **378**(1–2): pp. 124–129.

[17] Koh Y. K. K., Chiu T. Y., Boobis A. R., Cartmell E., Pollard S. J. T., Scrimshaw M. D., and Lester J. N., *A sensitive and robust method for the determination of alkylphenol polyethoxylates and their carboxylic acids and their transformation in a trickling filter wastewater treatment plant*. Chemosphere, 2008. **73**(4): pp. 551–556.

[18] Earls A., and Reydellet I., *Determination of specific alkylphenol ethoxylates in textiles*. 2006. [cited 31 March 2014] Available from: http://www.governmentchemist.org.uk/dm_documents/Determination%20of%20specific%20alkylphenol%20ethoxylates%20in%20textiles_6sQz8.pdf

[19] Motomlzu S., Fujiwara F., Fujiwara A., and Toei K., *Solvent extraction-spectrophotometric determination of anionic surfactants with ethyl violet*. Analysis Chemistry, 1982. **54**: pp. 392–397.

[20] Adak A., Pal A., and Bandyopadhyay M., *Spectrophotometric determination of anionic surfactants in wastewater using acridine orange*. Indian Journal of Chemical Technology, 2005. **12**: pp. 145–148.

[21] Guo P., Guan Z., Wang W., Chen B., and Huang Y., *Determination of linear alkylbenzene sulfonates by ion-pair solid-phase extraction and high-performance liquid chromatography*. Talanta, 2011. **84**: pp. 587–592.

[22] Villar M., Callejón M., Jiménez J. C., Alonso E., and Guiraúm A., *New rapid methods for determination of total LAS in sewage sludge by high performance liquid chromatography (HPLC) and capillary electrophoresis (CE)*. Analytica Chimica Acta, 2009. **634**: pp. 267–271.

[23] Bengoechea C., and Cantarero A. S., *Analysis of linear alkylbenzene sulfonate in waste water and sludge by high performance liquid chromatography: An exercise of validation*. Journal of Surfactants and Detergents, 2009. **12**(1): pp. 21–29.

[24] Villar M., Callejón M., Jiménez J. C., Alonso E., and Guiráum A., *Optimization and validation of a new method for analysis of linear alkylbenzene sulfonates in sewage sludge by liquid chromatography after microwave-assisted extraction*. Analytica Chimica Acta, 2007. **599**(1): pp. 92–97.

[25] McEvoy J., and Giger W., *Determination of linear alkylbenzenesulfonates in sewage sludge by high-resolution gas chromatography/mass spectrometry*. Environmental Science Technology, 1986. **20**: pp. 376–383.

[26] Kawauchi A., *Nonsolvent quantitation of anionic surfactants and inorganic ingredients in laundry detergent products*. Journal of the American Oil Chemists' Society, 1997. **74**(7): pp. 787–792.

[27] Frańska M., Ginter-Kramarczyk D., Szymański A., Kozik T., and Frański R., *Resistance of alkylphenol ethoxylate containing six ethoxylene units to biodegradation under the conditions of OECD (Organization for Economic Co-operation and Development) screening test*. International Biodeterioration & Biodegradation, 2009. **63**(8): pp. 1066–1069.

[28] Kiewiet A. T., Steen J. M. D. van der, and Parsons J. R., *Trace analysis of ethoxylated nonionic surfactants in samples of influent and effluent of sewage treatment plants by high-performance liquid chromatography*. Analysis Chemistry, 1995. **67**: pp. 4409–4415.

[29] Zhao J., Zhang G., Qin Y., and Zhao Y., *Aerobic biodegradation of alkylphenol ethoxylates*. Bioresource Technology, 2006. **97**(18): pp. 2478–2480.

[30] Cantero M., Rubio S., and Perez-Bendito D., *Determination of non-ionic polyethoxylated surfactants in sewage sludge by coacervative extraction and ion trap liquid chromatography-mass spectrometry*. Journal of Chromatogr A, 2004. **1046**(1–2): pp. 147–153.

[31] Marcomini A., Filipuzzi F., and Giger W., *Aromatic surfactants in laundry detergents and hard-surface cleaners: Linear alkylbenzenesulphonates and alkylphenol polyethoxylates*. Chemosphere, 1988. **17**(5): pp. 853–863.

[32] Winkler F. *The risks and alternatives of APEO and NPEO in textiles part 1*. 2013. [cited 23 September 2014] Available from: http://blog.stepchange-innovations.com/2013/11/risks-alternatives-apeo-npeo-in-textiles-1/.

[33] Hawrelak M., Bennett E., and Metcalfe C., *The environmental fate of the primary degradation products of alkylphenol ethoxylate surfactants in recycled paper sludge*. Chemosphere, 1999. **39**: pp. 745–752.

[34] Fernández-Sanjuan M., Rigol A., Sahuquillo A., Rodríguez-Cruz S., and Lacorte S., *Determination of alkylphenols and alkylphenol ethoxylates in sewage sludge: Effect of sample pre-treatment*. Analytical and Bioanalytical Chemistry, 2009. **394**(6): pp. 1525–1533.

[35] Scott M. J., and Jones M. N., *The biodegradation of surfactants in the environment*. Biochimica et Biophysica Acta (BBA): Biomembranes, 2000. **1508**(1–2): pp. 235–251.

[36] Fountoulakis M., Drillia P., Pakou C., Kampioti A., Stamatelatou K., and Lyberatos G. *Fate of nonylphenol and nonylphenol ethoxylates in sewage treatment plants*. In 9th International Conference on Environmental Science and Technology. 2005. Rhodes Island, Greece.

[37] Lee K. E., Morad N., Teng T. T., and Poh B. T., *Development, characterization and the application of hybrid materials in coagulation/flocculation of wastewater: A review*. Chemical Engineering Journal, 2012. **203**: pp. 370–386.

[38] vanLoon G. W., and Duffy S. J., *Enivirontal Chemistry, A Global Perpspective*. 2005, Oxford University Press Inc., New York, p. 515.

[39] Guida M., Mattei M., Della Rocca C., Melluso G., and Meriç S., *Optimization of alum-coagulation/flocculation for COD and TSS removal from five municipal wastewater*. Desalination, 2007. **211**(1–3): pp. 113–127.

[40] Lee C. S., Robinson J., and Chong M. F., *A review on application of flocculants in wastewater treatment*. Process Safety and Environmental Protection (2014), http://dx.doi.org/10.1016/j.psep.2014.04.010.

[41] Chan H., *High performance achieved by microbes to separate laundry effluents resulting in producing high water quality in a compact area*. Separation and Purification Technology, 2012. **90**: pp. 101–108.

[42] Wang C.-T., Chou W.-L., and Kuo Y.-M., *Removal of COD from laundry wastewater by electrocoagulation/electroflotation*. Journal of Hazardous Materials, 2009. **164**(1): pp. 81–86.

[43] Janpoor F., Torabian A., and Khatibikamal V., *Treatment of laundry waste-water by electrocoagulation*. Journal of Chemical Technology & Biotechnology, 2011. **86**(8): pp. 1113–1120.

[44] Önder E., Koparal A. S., and Öğütveren Ü. B., *An alternative method for the removal of surfactants from water: Electrochemical coagulation*. Separation and Purification Technology, 2007. **52**(3): pp. 527–532.

[45] Merzouk B., Yakoubi M., Zongo I., Leclerc J. P., Paternotte G., Pontvianne S., and Lapicque F., *Effect of modification of textile wastewater composition on electrocoagulation efficiency*. Desalination, 2011. **275**(1–3): pp. 181–186.

[46] Chithra K., Thilakavathi R., Murugan A. A., and Marimuthu C., *Treatment of textile effluent using sacrificial electrode*. Modern Applied Science, 2008. **2**: pp. 38–43.

[47] Dermentzis K., Christoforidis A., and Valsamidou E., *Removal of nickel, copper, zinc and chromium from synthetic and industrial wastewater by electrocoagulation*. International Journal of Environmental Sciences, 2011. **1**(5): pp. 697–710.

[48] Ramcharan T., and Bissessur A., *Treatment of laundry wastewater by biological and electrocoagulation methods*. Water Science & Technology, 2016. doi:10.2166/wst.2016.464.2016.

Section 4

Bioenergy for Resource Recovery

Chapter 11

Biohydrogen Production from Wastewaters

Periyasamy Sivagurunathan,

Gopalakrishnan Kumar, Arivalagan Pugazhendhi,

Guangyin Zhen, Takuro Kobayashi and Kaiqin Xu

Additional information is available at the end of the chapter

http://dx.doi.org/10.5772/65891

Abstract

Biohydrogen production technology is an emerging field for the advanced wastewater treatment with cogeneration of energy. Besides, hydrogen is an excellent candidate with high energy value (122 kJ/g) than other known carbon-based fuels with no adverse effects to the environment as it releases only water vapor as the by-products during the combustion. Biohydrogen production technology can be assisted through two major pathways: (a) light-dependent reaction (biophotolysis and photofermentation) and (b) light-independent reaction (dark fermentation and microbial electrohydrogenesis cells). The light-dependent reaction can be catalyzed by photosynthetic bacteria, whereas the dark fermentation catalyzed by the heterotrophic bacterial group of facultative and obligate anaerobes. The wastewaters are a rich source of organic nutrients which supports the growth of hydrogen producers along with the disposal of waste and energy recovery. In the present chapter, the recent advancements on biohydrogen production technology from wastewaters with respect to the (a) inoculum development, (b) process optimization, (c) scale-up and (d) the challenges and perspectives toward the improvement of this emerging technology for the wastewater treatment.

Keywords: biohydrogen, dark fermentation, wastewater

1. An overview of biohydrogen production

The growing demand of the energy for daily life purposes urged us to seek an alternative and renewable energy carrier with less emission of the pollutants. Hydrogen is an essential and promising candidate for replacing the fossil fuels depletion and greenhouse gas emission reduction. When burning, it releases only water vapor as a by-product with no adverse harmful gases such as NO_x and SiO_2, and hence, it is considered as clean and carbon-free energy carrier. The energy content of hydrogen is 122 kJ/g, which is 2.75-fold greater than the existing hydrocarbon fuels makes an ideal energy carrier for various industrial, transportation and power generations.

Different types of hydrogen production are available such as fossil fuel by hydrocarbon reforming, coal gasification and partial oxidation which requires high temperature and pressure. The biologically adopted hydrogen production methods can be classified as (i) biophotolysis of water using algae/cyanobacteria, (ii) photodecomposition of organic compounds using photosynthetic bacteria, (iii) dark fermentative hydrogen production using strict anaerobic or facultative bacteria and (iv) microbial fuel cells (MFC). Each biological production method had distinct advantages and limitations. For example, the green algae/cyanobacteria decomposes the water into gas (H_2) and liquid (H_2O) in the presence of sunlight by photosynthesis pathway, whereas the slow growth of the algal cells and an inhibition of hydrogenase enzyme with the presence of traces of oxygen limit their application in large scale extent. The photosynthetic bacteria and dark fermentation bacteria share a similar metabolism for the breakdown of organic compounds for their energy and the liberation of energy [1, 2]. The photosynthetic bacteria use organic acids as a substrate and prone to the ammonium and oxygen toxicity, making it as unsuitable for commercial hydrogen production. In contrast, the dark fermentation degrades wide range of organic waste from complex lignocellulose, food waste and industrial wastewater to simpler monomers (sucrose, glucose). However, the chemical oxygen demand (COD) removal efficiency of the dark fermentation is relatively lower 33%, as it requires further treatment before discharge into the system. Moreover, the biomass growth rate and hydrogen production rate of the dark fermentation are comparatively higher than the other hydrogen production methods and make it as attractive candidate for industrial and commercial biohydrogen production [3]. Recently, the auxiliary methods for the hydrogen production from hydrogen effluent have been emerged through microbial fuel cell (MFC) or bioelectrochemical systems (BES) technology.

2. Hydrogen-producing microorganisms

Table 1 displayed the microbial strains helpful for biohydrogen production through dark fermentation [4]. Hydrogen production during fermentation involves either facultative anaerobic bacteria or strict anaerobic bacteria. Facultative anaerobes are capable of growing in the absence of oxygen. The most common hydrogen-*producing* facultative anaerobes are *Klebsiella pneumoniae* [5], *Escherichia coli* [6], *Enterobacter aerogenes* [7], *Rhodospirillum rubrum, Methanobacterium formiccium* [4]. Chookaew et al. [5] reported that *Klebsiella* sp. TR17 is able to produce biohydrogen from crude glycerol in an up-flow anaerobic sludge blanket (UASB) reactor with highest HPR of 242.15 mmol H_2/L/d and HY of 44.27 mmol H_2/g glycerol. Besides, the *Klebsiella pneumoniae* produce valuable by-products such as 1,3-propanediol and 2,3-butanediol [8]. Reungsang et al. [7] reported that the immobilized *E. aerogenes* ATCC

Wastewater type	Inoculum source	Hydrogen yield (HY) (mol/mol hexose added)	References
Distillery effluent	*Enterobacter cloacae*	165.3 mL/g COD	[32]
Cassava WW	*Clostridium acetobutylicum*	2.41 mol/mol glu	[33]
Rice mill WW	*Enterobacter aerogens*	1.74 mol/mol sugar	[34]
Rice mill WW	*Citrobacter ferundii*	1.40 mol/mol sugar	[34]
Rice mill WW	*Enterobacter aerogens* RM08	1.97 mol/mol	[35]
CMS	*Clostridium tyrobutyricum*	0.7 mmol H_2/g COD	[36]
CMS	*Clostridium pasteurianum*	1.1 mmol H_2/g COD	[36]
CMS	*Clostridium sporosphaeroides*	0.9 mmol H_2/g COD	[36]

Table 1. Hydrogen production using pure cultures WW, wastewater; CMS, condensed molasses soluble.

13048 produced major soluble metabolite products (SMPs), such as ethanol, 1,3-propanediol (1,3-PD), formic acid and acetic acid.

2.1. Facultative anaerobes

Facultative anaerobes play important roles in H_2 production by biological routes, as it can grow in the presence of oxygen, higher biomass growth rate and utilization of wide range of organic wastes. The widely studied facultative anaerobic model for hydrogen production is *E. coli* and *E. aerogenes*. Facultative anaerobes convert pyruvate to acetyl-coA and formate with the catalysis of pyruvate formate-lyase complex and then release H_2 with formate hydrogen lyase. The maximum theoretical hydrogen yield is 2 mol of H_2 per mole of glucose. The glucose metabolic pathway yields succinate, lactate, acetate, ethanol and formate, as fermentation end-products. *Enterobacter* sp. have been widely used in various reactor configuration from batch to continuous mode operation. Several attempts like coculture of the facultative anaerobes with strict anaerobes have been assessed to improve the biohydrogen production. The coculture has advantages over pure culture due to the less maintenance, technical feasibility and faster substrate utilization rate. Sivagurunathan et al. [9] demonstrated that the addition of enriched mixed culture with *Enterobacter cloacae* enhanced the hydrogen production rate of 2.25 L/L-d from beverage wastewater. In another report [6], immobilization of *E. coli* cells using sodium alginate increased the hydrogen production efficiency from fructose (1.17 mol/mol hexose) and beverage wastewater (1.65 mol/mol hexose), respectively.

2.2. Mixed consortia

The mixed consortia can be derived from a variety of different natural sources, such as sewage sludge, anaerobically digested sludge, compost, animal manure and contaminated soil (**Table 2**). Mixed culture contains different types of bacteria; it also contains methanogens or hydrogen-consuming bacteria. Mixing also determines the local shear stress that the flow applies to microorganisms. Mixed culture can be obtained from aerobic or anaerobic sludge in wastewater treatment plants or compost piles or any other source of bacteria. Currently,

Wastewater type	Inoculum source	Hydrogen yield (HY) (mol/mol hexose added)	References
BWW	EMC-sewage sludge + pig slurry	1.95 mol/mol glu	[37]
BWW	EMC + *E. coli* XL1 blue	260 mL/g COD	[11]
Sugar beet juice	Anaerobic sludge	2.0 mol/mol glu	[38]
Distillery WW	Anaerobic sludge	10.95 mmol/g COD	[39]
Dairy WW	Anaerobic sludge	15.33 mmol/g COD	[40]
Cheese processing WW	Mixed cultures	10.2 mM/g COD	[41]
Organic WW	Soil	2.32 mol/mol	[42]
Herbal WW	Slaughter house sludge	165 mL/g COD	[43]
CMS	Anaerobic sludge	1.5 mol/mol	[44]
Brewery WW	Anaerobic sludge	1.21 mol/mol	[45]
GWW	Anaerobic sludge	0.75 mol/mol	[46]

WW, wastewater; BWW, beverage wastewater; CMS, condensed soluble molasses; GWW, glycerine wastewater.

Table 2. Hydrogen production using mixed consortia.

researchers mainly focused two routes for microbial fermentative hydrogen production: one utilizes pure microbial strains and the other employs a mixed microbial consortium. Generally, the hydrogen-producing efficiency and hydrogen yield of pure bacteria are lower than mixed consortia. Several investigators have focused on hydrogen production by microbial fermentation using a mixed microbial consortium, because of low-cost organic substrates, high hydrogen yields and operated in non-sterile conditions.

3. Process optimization for scale-up

Biohydrogen production is an emerging research area in the sustainable biofuel production via anaerobic fermentation technology. Though the hydrogen production from biological routes seems attractive over other commercial process, the operational conditions are essential to optimize in order to attain the maximum achievable hydrogen production rates and yields. A few important parameters on these aspects are as follows:

(a) Inoculum pretreatment
(b) pH
(c) Nutrient availability
(d) Hydraulic retention time

Biohydrogen production through mixed consortia is a complex bioprocess where the inoculum source, substrate type, environmental factors (pH, temperature and substrate concentration), nutrient availability and HRT can influence the metabolic reactions of hydrogen producers. Optimizing these factors is a paramount importance for enhancing the hydrogen production efficiency from organic wastes.

3.1. Inoculum pretreatment

The active acidogenic hydrogen-producing biocatalyst role is crucial, notably in a complex mixed culture microenvironment. In general, the hampering hydrogen yield from mixed consortia was observed due to (i) the competition of hydrogen-consuming microbes and (ii) diversion of the metabolic flux toward non-favorable hydrogen by-products. The hydrogen consumers, such as lactic acid bacteria, methanogenes and sulfur-reducing bacteria, not only act as a competitor for the hydrogen producers but also synthesize various by-products, which affect the growth of hydrogen producers. For instance, the release of proteinaceous toxin (bacteriocins) by lactate-producing bacteria acts as a suppressing factor for hydrogen production and microbial growth [10]. Thus, when the mixed culture is used as an inoculum source, pretreatment step acts as an important role in determining the efficiency of the hydrogen production from mixed consortia. **Table 3** showed the various pretreatment methods for enriching the hydrogen producers. The pretreatment step promotes the selective enrichment of hydrogen producers with a suppression of the hydrogenotrophic methanogenes and other hydrogen consumers. The suppression of the hydrogen consumers by pretreatment process allows the mixed consortia to produce the hydrogen as a major product. The fundamental basics relied with the pretreatment method are the physiological difference of the microorganisms. The spore-forming hydrogen producers survive under the harsh pretreatment conditions, whereas the vegetative cells ruptured/killed during the pretreatment. Various pretreatment methods, such as heat shock, acid shock, alkali shock, chemical agents, load shock and oxygen shock, have been assessed for enriching the hydrogen producers from mixed consortia. Each pretreatment step has a significant impact on the suppression of the microbial populations and also the distribution of the microbial metabolism.

Among the various pretreatment methods, the heat-shock [11] pretreatment has been widely accepted as a suitable method for preparing the hydrogen-producing seed inocula, due to the relatively simple method for the suppression of the hydrogen consumers and selective enrich-

Substrate	Inoculum source	Pretreatment method	Hydrogen yield (HY)	References
Deoiled jatropha waste	Anaerobic digester sludge	Heat shock	20 mL H_2/g VS	[11]
Glucose	Anaerobic sludge	Acid shock	0.80 mol/mol	[12]
Sucrose	Anaerobic digester sludge	Base shock	3.06 mol/mol	[13]
Glucose	Anaerobic granular sludge	Chloroform	1.55 mol/mol	[14]
Desugared molasses	Digested manure	Load shock	237 mL H_2/g- sugar	[16]
Glucose	Anaerobic sludge	Repeated aeration	1.96 mol/mol	[17]
Glucose	Anaerobic sludge	Gamma irradiation	2.15 mol/mol	[47]

Table 3. Inoculum pretreatment method for enriching hydrogen production mixed consortia.

ment of the sporulating hydrogen-producing bacteria such as *Clostridium* sp. The acid-shock [12] and base-shock [13] pretreatments suppress the methanogenic activity by the narrow selective growth pH range of the methanogenes (6–7.5), whereas the *Clostridium* populations survive in the harsh condition due to the spore-forming capability. The chemical shock methods such as chloroform [14] and 2-bromoethanesulfonic acid (BESA) [15] have a complex structure, analog to the methanogenic coenzyme, and it acts as a inhibitor for the methanogenes. This method facilitated the suppression of the methanogenes, whereas the other non-spore-forming hydrogen producers such as *Enterobacter* sp. can also survive with the presence of *Clostridium* sp., thus enhancing the substrate utilization and hydrogen yield. The load-shock [16] treatment is directed by the exposure of the inoculum to a higher substrate concentration, and it leads to the surge in the pH with an accumulation of organic acids and inhibits the methanogenic populations.

Ren et al. [17] demonstrated that application of various pretreatment methods, such as acid, alkaline, heat-shock and repeated aeration, can greatly affect the metabolic pathway and the microbial community distribution pattern. The dominant butyric acid-mediated hydrogen metabolism was observed with heat-shock and alkaline treatment, and mixed-type fermentation pathway was observed with the acid pretreatment, whereas the ethanol-type pathway was observed with repeated aeration treatment with a maximum hydrogen yield of 1.96 mol/mol glucose. The microbial community characterized by denaturing gradient gel electrophoresis (DGGE) revealed that the changes in the composition of the microbial dynamics affect the hydrogen yield. The strain *Ethanoligenens harbinens* was detected under repeated aeration condition with an ethanol-mediated pathway, and the hydrogen-consuming propionic acid bacterium *Propionibacterium propionicus* was detected in acid treatment with low hydrogen productivity. The heat-shock-mediated mixed culture was dominated with *Clostridium* sp. which represents the butyric-acid-type metabolic pathway. Based on the evidence, the appropriate pretreatment method is essential for enriching the hydrogen-producing bacterial populations and enhanced hydrogen production.

3.2. pH

pH is the key driven parameter affecting the cellular metabolism of hydrogen-producing bacterial populations, since the prevalent end products of the bacterial metabolism vary with the changes in the medium pH. Based on the pH and the major end products formation, three metabolic pathways have been proposed (a) ethanol type (EtOH) (Eq. 1), (b) butyric type (HBu) (Eq. 2) and (c) propionic type (HPr) (Eq. 3). The former, HBu type, involved in the hydrogen-generating reactions, whereas the latter, HPr type, involved in the hydrogen-scavenging reactions. Hence, the elimination of the propionate formation is an essential step for the enhancement of hydrogen production.

$$C_6H_{12}O_6 + 2\,H_2O \rightarrow 2\,CH_3CH_2OH + 2\,HCO_3^- + 2\,H_2 \tag{1}$$

$$\Delta G^0 = -235.0\ \text{kJ/mol}$$

$$C_6H_{12}O_6 \rightarrow CH_3CH_2CH_2COOH + 2\,CO_2 + 2\,H_2 \quad (2)$$

$$\Delta G_0' = -254.0\ \text{kJ/mol}$$

$$C_6H_{12}O_6 + 2\,H_2 \rightarrow 2\,CH_3CH_2COOH + 2\,H_2O \quad (3)$$

$$\Delta G_0' = -279.4\ \text{kJ/mol}$$

pH affects the physiological conditions of the bacterial growth, metabolism and ions transport. Optimizing the pH is considering a key factor influenced the redox environment and the direction of electron flow toward the hydrogen formation. The experimental reports demonstrated that the optimal pH for the bacterial growth does not result in the elevated hydrogen production performances [3]. For the dark fermentative hydrogen fermentation, the optimal pH for efficient hydrogen production lied between 5.5 and 6.5 for various wastewaters and pure substrates [18]. In addition, the acidic pH induces the pyruvate transformation to volatile fatty acids (VFA) with concomitant hydrogen production, whereas the neutral pH facilitated the methanogenic pathway. Maintaining the acidogenic (5.5–6.5) pH is essential for controlling the methanogenic populations and efficient hydrogen production.

3.3. Nutrients

The inorganic nutrient supplements, such as nitrogen (N), phosphorus (P) and iron (Fe), along with carbon (C) source, are important for microbial growth and improvement in the hydrogen production. The nutrient at proper concentration is beneficial for hydrogen production. For instance, Lin and Lay [19] explained that at a carbon/nitrogen (C/N) ratio of 47, the hydrogen yield from sucrose was 1.9 times higher than the control with a value of 4.8 mol/mol substrate. In a pure culture thermotolerant *Kelbsiella* sp., the maximum hydrogen yield of 0.28 mol/mol glycerol was observed with 11.21 g/L glycerol, 2.84 g/L KH_2PO_4 and 5.66 g/L NH_4Cl, respectively [20]. Wang et al. [21] mentioned that the hydrogen production efficiency of glucose (313.3 mL/g glucose) was improved with low supplementation of nitrate 0.1 g/L; however, increased concentration of nitrate over 0.1 g/L significantly affected the hydrogen yield and the substrate consumption rate. The drop in hydrogen production is attributed by the inhibition of nitrogenize activity by surplus ammonium ions [22, 23]. The iron (Fe) is an important element essential for the hydrogenase activity, which directs the metabolic pathway by stimulating the active site for the ferredoxin (Fd). The addition of iron supplement was shown to improve the hydrogen production. Gadhe et al. [24] demonstrated the effects of nano-sized iron and nickel oxide nanoparticles by using dairy wastewater as a substrate, and it showed that an enhancement in hydrogen yield of 17.2 mmol/g COD is due to the enhanced activity of the ferredoxin oxidoreductase, ferredoxin and hydrogenase enzymes. Moreover, the optimal value for the Fe^{2+} concentration is varied with the type of substrates used. For instance, the optimal concentration reported by Liu and Shen [25] was 10 mg/L from starch, whereas palm oil mill effluent showed an optimal value of 257 mg/L [26].

3.4. Hydraulic retention times

The hydraulic retention time (HRT) is one of the key process control parameters influencing the continuous hydrogen production. HRT enables the better process control of the microorganisms that can regulate the metabolic pathway favorable for efficient hydrogen production. The long HRT permits the growth of hydrogen consumers mainly archaea, which is unsuitable for hydrogen production, whereas too low HRT leads to the washout of active biomass and deterioration of the reactor performances. The optimization of HRT is a paramount importance for the scale-up, long-term and sustainable hydrogen production. HRT controls the organic loading rate (OLR), substrate degradation and reaction kinetics. The organic wastes required long HRT, whereas the simple organics required short HRT [2]. The reported optimum HRT value for the wastewater ranges from 0.5 to 24 h. For example, the short HRT (0.5 h) provided the maximum hydrogen production rate of 14 L/L-d from condensed soluble wastewater [27], whereas the long HRT (24 h) is required for efficient conversion of olive mill wastewater with a HPR of 7.0 L/L-d [28]. The process parameters discussed above significantly influenced the hydrogen production; hence, careful assessment of each individual factor is important for stable hydrogen production.

4. Bioreactor design considerations for continuous hydrogen production

Bioreactor configuration is a notable factor in dark fermentative hydrogen production, as it influences the contact between the organic waste and hydrogen producers, substrate utilization, biomass dilution rate, etc. According to the feeding regime, the biohydrogen production can be conducted in batch, semi-continuous and continuous mode (**Table 4**). The batch mode operation is relatively simple and easier to control. Hence, the batch mode hydrogen reactors have been widely used to determine the feasibility of the organic waste feedstock and to optimize the environmental parameters such as pH, temperature, substrate concentration. In semi-continuous mode operation, the organic substrate was operated in a sequencing batch which includes feeding, reaction, settle and decant stages [29]. The sequencing batch operation is recommended for a viscous substrate like a POME and solid organic biomass like food waste and lignocellulosic biomass, where the physical contact between the substrate and microorganisms is limited, and this reactor mode operation enables the better hydrolysis rate, avoids clogging in the pipes and retains the effective biomass concentration. In continuous mode operation, the continuous supply of nutrients and the removal of the pollutants occur simultaneously with the aid of peristaltic pumps.

Although various reactor models assessed, the continuous mode operation is preferred for bench-scale and commercial-scale applications. The widely investigated model for continuous mode operation is the CSTR type, wherein the substrates and feedstocks are well mixed inside the reactor with the aid of the mechanical rotor; however, the biomass washout usually occurred at lower HRT [27, 30]. In some cases, the biofilm formed inside the CSTR is resistance to the biomass washout and thereby enhancing the hydrogen production performances. Chu et al. [27] investigated the CSTR reactor model by using condensed soluble molasses as a substrate with suspended and immobilized cells as inoculum source. The hydrogen production

Substrate	Inoculum source	Reactor mode	HPR (L/L/d)	References
Palm oil mill effluent	Anaerobic digester sludge	ASBR	6.7	[29]
Condensed molasses	Anaerobic sludge	CSTR	14.04	[27]
Beverage WW	Enriched mixed cultures	CSTR	37.56	[30]
Tofu processing WW	Anaerobic digester sludge	MBR	19.86	[48]
Desugared molasses	Anaerobic sludge	UASB	5.6	[49]
Olive mill WW	Anaerobic sludge	PBR	7.0	[28]
Beverage WW	Enriched mixed cultures	ICBR	55.4	[31]
Beverage WW	Anaerobic digester sludge	PBR	88.7	[50]

WW, wastewater; ASBR, anaerobic sequencing batch reactor; CSTR, continuously stirred tank reactor; MBR, membrane bioreactor; PBR, packed bed reactor; ICBR, immobilized cell bioreactor.

Table 4. Bioreactor types used in hydrogen production.

from immobilized cell was relatively lower with a maximum HPR of 7.6 L/L/d; however, the suspended cells operation provided the maxim HPR of 14.04 L/L/d, respectively. The observed variation is attributed by the washout of the active biomass in immobilized cells system (9.8 g volatile suspended solids (VSS)/L), poor mass transfer between the microbes and substrates and the increased lactic acid formation. On the other hand, the suspended cell system formed a hydrogen-producing granule (HPG) inside the reactor, and thus, it retains the active biomass (12.30 g VSS/L) and less formation of the lactic acid. Sivagurunathan et al. [30] demonstrated that the hydrogen production from ICBR [31] was higher (55 L/L/d) than the suspended cells CSTR (37.56 L/L/d) operation. The superior performance of the ICBR is due to the formation of granular biomass at short HRT of 3 h with the presence of *Selenomonas* sp. and further maturation of granules with the presence of active hydrogen-producing *Clostridium* Sp. The *Selenomonas* sp. act as a bio-glue for the development of granules. Moreover, the energy content analysis of the beverage wastewater with immobilized cells system analysis showed that it has the capability of reducing the CO_2 reduction efficiency of 2832 ton CO_2 equivalent/year.

5. Conclusion

Biohydrogen production from industrial wastewaters seems to be appropriate and environmental benign option for future sustainable hydrogen economy with simultaneous energy recovery and waste disposal. Various studies revealed the hydrogen production potential of wastewaters. Among them, sugar-rich wastewaters are the promising substrate for high-efficient hydrogen production rates and yields, due to their easier degradation rate and higher substrate concentration. Other key challenges that rely on dark fermentative hydrogen production from organic wastes are the low substrate conversion efficiency, moderate-to-low hydrogen yield and residual organics in the effluents. In general, biohydrogen production is a primary step for wastewater treatment, in which a maximum 4 mol/mol glucose representing 33% of COD removal efficiency; nearly 70–80% of the residual organics remain untreated with the hydrogen-producing effluent, thus seeks further disposal of the effluent in the wastewater streams. The post-residual effluent has to be integrated with various two-step processes, such as methane production, photofermentation, microbial electrolysis cells, bioplastics production and microalgae cultivation, for maximizing the energy recovery.

The author Dr. Periyasamy Sivagurunathan greatly acknowledged the financial assistance from Japan Society for Promotion of Science: JSPS ID: P15370 for this study.

Author details

Periyasamy Sivagurunathan[1]*, Gopalakrishnan Kumar[1], Arivalagan Pugazhendhi[2], Guangyin Zhen[3], Takuro Kobayashi[1] and Kaiqin Xu[1, 4]*

*Address all correspondence to: contact2sivas12@gmail.com and joexu@nies.go.jp

1 Center for Materials Cycles and Waste Management Research, National Institute for Environmental Studies, Tsukuba, Japan

2 Department of Environmental Engineering, Daegu University, Gyeongsan, Gyeongbuk, Republic of Korea

3 School of Ecological and Environmental Sciences, East China Normal University, Shanghai, China

4 School of Environmental Science and Engineering, Shanghai Jiao Tong University, Shanghai, China

References

[1] Singh L, Wahid ZA. Methods for enhancing bio-hydrogen production from biological process: A review. Journal of Industrial and Engineering Chemistry. 2015;21:70–80.

[2] Sivagurunathan P, Kumar G, Bakonyi P, Kim S-H, Kobayashi T, Xu KQ, et al. A critical review on issues and overcoming strategies for the enhancement of dark fermentative hydrogen production in continuous systems. International Journal of Hydrogen Energy. 2016;41:3820–36.

[3] Ghimire A, Frunzo L, Pirozzi F, Trably E, Escudie R, Lens PNL, et al. A review on dark fermentative biohydrogen production from organic biomass: Process parameters and use of by-products. Applied Energy. 2015;144:73–95.

[4] Elsharnouby O, Hafez H, Nakhla G, El Naggar MH. A critical literature review on biohydrogen production by pure cultures. International Journal of Hydrogen Energy. 2013;38:4945–66.

[5] Chookaew T, O-Thong S, Prasertsan P. Biohydrogen production from crude glycerol by immobilized *Klebsiella* sp. TR17 in a UASB reactor and bacterial quantification under non-sterile conditions. International Journal of Hydrogen Energy. 2014;39:9580–7.

[6] Sivagurunathan P, Kumar G, Lin C-Y. Hydrogen and ethanol fermentation of various carbon sources by immobilized *Escherichia coli* (XL1-Blue). International Journal of Hydrogen Energy. 2014;39:6881–8.

[7] Reungsang A, Sittijunda S, Angelidaki I. Simultaneous production of hydrogen and ethanol from waste glycerol by *Enterobacter aerogenes* KKU-S1. International Journal of Hydrogen Energy. 2013;38:1813–25.

[8] Chookaew T, O-Thong S, Prasertsan P. Fermentative production of hydrogen and soluble metabolites from crude glycerol of biodiesel plant by the newly isolated thermotolerant *Klebsiella pneumoniae* TR17. International Journal of Hydrogen Energy. 2012;37:13314–22.

[9] Sivagurunathan P, Gopalakrishnan K, Lin CY. Enhancement of fermentative hydrogen production from beverage wastewater via bioaugmentation and statistical optimization. Current Biochemical Engineering. 2014;1:92–8.

[10] Gomes SD, Fuess LT, Mañunga T, Feitosa de Lima Gomes PC, Zaiat M. Bacteriocins of lactic acid bacteria as a hindering factor for biohydrogen production from cassava flour wastewater in a continuous multiple tube reactor. International Journal of Hydrogen Energy. 2016;41:8120–31.

[11] Kumar G, Bakonyi P, Sivagurunathan P, Nemestóthy N, Bélafi-Bakó K, Lin C-Y. Improved microbial conversion of de-oiled Jatropha waste into biohydrogen via inoculum pretreatment: process optimization by experimental design approach. Biofuel Research Journal. 2015;2:209–2014.

[12] Wang J, Wan W. Comparison of different pretreatment methods for enriching hydrogen-producing bacteria from digested sludge. International Journal of Hydrogen Energy. 2008;33:2934–41.

[13] Zhu H, Béland M. Evaluation of alternative methods of preparing hydrogen producing seeds from digested wastewater sludge. International Journal of Hydrogen Energy. 2006;31:1980–8.

[14] Hu B, Chen S. Pretreatment of methanogenic granules for immobilized hydrogen fermentation. International Journal of Hydrogen Energy. 2007;32:3266–73.

[15] Chang S, Li J-Z, Liu F. Evaluation of different pretreatment methods for preparing hydrogen-producing seed inocula from waste activated sludge. Renewable Energy. 2011;36:1517–22.

[16] Kongjan P, O-Thong S, Angelidaki I. Biohydrogen production from desugared molasses (DM) using thermophilic mixed cultures immobilized on heat treated anaerobic sludge granules. International Journal of Hydrogen Energy. 2011;36:14261–9.

[17] Ren N-Q, Guo W-Q, Wang X-J, Xiang W-S, Liu B-F, Wang X-Z, et al. Effects of different pretreatment methods on fermentation types and dominant bacteria for hydrogen production. International Journal of Hydrogen Energy. 2008;33:4318–24.

[18] Lin C-Y, Lay C-H, Sen B, Chu C-Y, Kumar G, Chen C-C, et al. Fermentative hydrogen production from wastewaters: A review and prognosis. International Journal of Hydrogen Energy. 2012;37:15632–42.

[19] Lin CY, Lay CH. Effects of carbonate and phosphate concentrations on hydrogen production using anaerobic sewage sludge microflora. International Journal of Hydrogen Energy. 2004;29:275–81.

[20] Chookaew T, O-Thong S, Prasertsan P. Statistical optimization of medium components affecting simultaneous fermentative hydrogen and ethanol production from crude glycerol by thermotolerant *Klebsiella* sp. TR17. International Journal of Hydrogen Energy. 2014;39:751–60.

[21] Wang B, Wan W, Wang J. Effects of nitrate concentration on biological hydrogen production by mixed cultures. Frontiers of Environmental Science & Engineering in China. 2009;3:380–6.

[22] Wang B, Wan W, Wang J. Effect of ammonia concentration on fermentative hydrogen production by mixed cultures. Bioresource Technology. 2009;100:1211–3.

[23] Redwood MD, Macaskie LE. A two-stage, two-organism process for biohydrogen from glucose. International Journal of Hydrogen Energy. 2006;31:1514–21.

[24] Gadhe A, Sonawane SS, Varma MN. Enhancement effect of hematite and nickel nanoparticles on biohydrogen production from dairy wastewater. International Journal of Hydrogen Energy. 2015;40:4502–11.

[25] Liu G, Shen J. Effects of culture and medium conditions on hydrogen production from starch using anaerobic bacteria. Journal of Bioscience and Bioengineering. 2004;98:251–6.

[26] O-Thong S, Prasertsan P, Intrasungkha N, Dhamwichukorn S, Birkeland N-K. Optimization of simultaneous thermophilic fermentative hydrogen production and COD reduction from palm oil mill effluent by Thermoanaerobacterium-rich sludge. International Journal of Hydrogen Energy. 2008;33:1221–31.

[27] Chu CY, Wu SY, Hsieh PC, Lin CY. Biohydrogen production from immobilized cells and suspended sludge systems with condensed molasses fermentation solubles. International Journal of Hydrogen Energy. 2011;36:14078–85.

[28] Scoma A, Bertin L, Fava F. Effect of hydraulic retention time on biohydrogen and volatile fatty acids production during acidogenic digestion of dephenolized olive mill wastewaters. Biomass and Bioenergy. 2013;48:51–8.

[29] Badiei M, Jahim JM, Anuar N, Sheikh Abdullah SR. Effect of hydraulic retention time on biohydrogen production from palm oil mill effluent in anaerobic sequencing batch reactor. International Journal of Hydrogen Energy. 2011;36:5912–9.

[30] Sivagurunathan P, Lin C-Y. Enhanced biohydrogen production from beverage wastewater: process performance during various hydraulic retention times and their microbial insights. RSC Advances. 2016;6:4160–9.

[31] Sivagurunathan P, Sen B, Lin C-Y. High-rate fermentative hydrogen production from beverage wastewater. Applied Energy. 2015;147:1–9.

[32] Mishra P, Das D. Biohydrogen production from Enterobacter cloacae IIT-BT 08 using distillery effluent. International Journal of Hydrogen Energy. 2014;39:7496–507.

[33] Cappelletti BM, Reginatto V, Amante ER, Antônio RV. Fermentative production of hydrogen from cassava processing wastewater by *Clostridium acetobutylicum*. Renewable Energy. 2011;36:3367–72.

[34] Ramprakash B, Muthukumar K. Comparative study on the production of biohydrogen from rice mill wastewater. International Journal of Hydrogen Energy. 2014;39:14613–21.

[35] Ramprakash B, Muthukumar K. Comparative study on the performance of various pretreatment and hydrolysis methods for the production of biohydrogen using *Enterobacter aerogenes* RM 08 from rice mill wastewater. International Journal of Hydrogen Energy. 2015;40:9106–12.

[36] Hsiao C-L, Chang J-J, Wu J-H, Chin W-C, Wen F-S, Huang C-C, et al. Clostridium strain co-cultures for biohydrogen production enhancement from condensed molasses fermentation solubles. International Journal of Hydrogen Energy. 2009;34:7173–81.

[37] Sivagurunathan P, Sen B, Lin C-Y. Batch fermentative hydrogen production by enriched mixed culture: Combination strategy and their microbial composition. Journal of Bioscience and Bioengineering. 2014;117:222–8.

[38] Dhar BR, Elbeshbishy E, Hafez H, Lee H-S. Hydrogen production from sugar beet juice using an integrated biohydrogen process of dark fermentation and microbial electrolysis cell. Bioresource Technology. 2015;198:223–30.

[39] Gadhe A, Sonawane SS, Varma MN. Evaluation of ultrasonication as a treatment strategy for enhancement of biohydrogen production from complex distillery wastewater and process optimization. International Journal of Hydrogen Energy. 2014;39:10041–50.

[40] Gadhe A, Sonawane SS, Varma MN. Enhanced biohydrogen production from dark fermentation of complex dairy wastewater by sonolysis. International Journal of Hydrogen Energy. 2015;40:9942–51.

[41] Yang P, Zhang R, McGarvey JA, Benemann JR. Biohydrogen production from cheese processing wastewater by anaerobic fermentation using mixed microbial communities. International Journal of Hydrogen Energy. 2007;32:4761–71.

[42] Sharma Y, Li B. Optimizing hydrogen production from organic wastewater treatment in batch reactors through experimental and kinetic analysis. International Journal of Hydrogen Energy. 2009;34:6171–80.

[43] Sivaramakrishna D, Sreekanth D, Sivaramakrishnan M, Sathish Kumar B, Himabindu V, Narasu ML. Effect of system optimizing conditions on biohydrogen production from herbal wastewater by slaughterhouse sludge. International Journal of Hydrogen Energy. 2014;39:7526–33.

[44] Wu J-H, Lin C-Y. Biohydrogen production by mesophilic fermentation of food wastewater. Water Science and Technology. 2004;49:223–8.

[45] Shi X-Y, Jin D-W, Sun Q-Y, Li W-W. Optimization of conditions for hydrogen production from brewery wastewater by anaerobic sludge using desirability function approach. Renewable Energy. 2010;35:1493–8.

[46] Fernandes BS, Peixoto G, Albrecht FR, Saavedra del Aguila NK, Zaiat M. Potential to produce biohydrogen from various wastewaters. Energy for Sustainable Development. 2010;14:143–8.

[47] Yin Y, Hu J, Wang J. Enriching hydrogen-producing bacteria from digested sludge by different pretreatment methods. International Journal of Hydrogen Energy. 2014;39:13550–6.

[48] Kim MS, Lee DY, Kim DH. Continuous hydrogen production from tofu processing waste using anaerobic mixed microflora under thermophilic conditions. International Journal of Hydrogen Energy. 2011;36:8712–8.

[49] Kongjan P, O-Thong S, Angelidaki I. Hydrogen and methane production from desugared molasses using a two-stage thermophilic anaerobic process. Engineering in Life Sciences. 2013;13:118–25.

[50] Chu C-Y, Hastuti ZD, Dewi EL, Purwanto WW, Priyanto U. Enhancing strategy on renewable hydrogen production in a continuous bioreactor with packed biofilter from sugary wastewater. International Journal of Hydrogen Energy. 2016;41:4404–12.

Chapter 12

Valorization of Glucose-Based Wastewater Through Production of Hydrogen, Volatile Fatty Acids and Alcohols

Eduardo Lucena Cavalcante de Amorim,
Leandro Takano Sader, Lucas Rodrigues Ramos and
Edson Luiz Silva

Additional information is available at the end of the chapter

http://dx.doi.org/10.5772/67101

Abstract

The production of hydrogen in an anaerobic fluidized bed reactor (AFBR) was evaluated under different organic loading rates (OLRs) with the addition of 1 g L^{-1} sodium bicarbonate for pH control. Expanded clay was used as the support material for microbial attachment. Two AFBRs were operated with glucose concentrations of 10 and 25 g L^{-1} and a hydraulic retention time (HRT) decreasing from 8 to 1 h at a controlled temperature of 30°C. A linear correlation was observed between the hydrogen production rate (HPR) and the OLR, except for the reactor operated with 25 g L^{-1} glucose. The maximum HPR of 1.58 L h^{-1} L^{-1} was obtained with an HRT of 1 h, and the maximum H_2 yield of 1.32 mol H_2 mol^{-1} glucose was obtained with an HRT of 2 h, in the reactor operated with 10 g L^{-1} glucose.

Keywords: hydrogen production, anaerobic fluidized bed reactor, substrate concentration, hydraulic retention time, organic loading rate

1. Introduction

The acidogenic fermentation of wastewater or biowaste for H_2 production has attracted great global interest because it is a cheap and simple technology that produces clean energy from renewable sources while reducing pollutants [1, 2].

According to Reddy et al. [3], one of the major drawbacks of using organic wastes is that only 30–40% of the substrate is used to H_2 production and 60–70% is converted to several other metabolites. However, some metabolites are commercially attractive, such as acetic acid, butyric acid, propionic acid, lactic acid, succinic acid, 1,3-propanediol, ethanol, methanol, etc. [4, 5].

H_2 production has been carried out with a variety of organic wastes, in which the source of carbonaceous organic material is based on glucose, sucrose, starch, xylose, cheese-processing wastewater, tapioca-processing wastewater, and sugarcane vinasse [6–9].

The fermentation process for the production of H_2 in anaerobic reactors is greatly influenced by several factors, such as the type of wastewater, the inoculum, the type of reactor, the nutritional requirements, the temperature, and the pH [10–12].

For practical engineering, industrial H_2 production requires continuous or semicontinuous production processes. Several types of reactors have been studied to effectively generate H_2. Reactors for continuous H_2 production include suspended biomass reactors, e.g., continuous stirred tank reactors (CSTRs) [13–15] and anaerobic sequencing bed reactors (ASBRs) [16], and biofilm reactors such as anaerobic packed bed reactors (APBRs) [17] and anaerobic fluidized bed reactors (AFBRs) [6–9, 18]. The advantages and disadvantages of different reactor types vary. Biofilm reactors can overcome the drawbacks of suspended biomass reactors by decoupling the biomass retention time from HRT, thus increasing the biomass concentration in the reactor. The hydraulic mixing regime is usually more turbulent in AFBRs than in APBRs, which improves mass transfer and treatment efficiencies because bed fluidization favors contact between the biofilm and substrate [19–21].

Hydrogen production is a microbial-mediated process dependent on several parameters that can affect the performance. Some of these are the inoculum source, pH, substrate concentration, accessible nutrients, HRT, and temperature [11, 21]. Their control in appropriate range can enrich the microbial community with hydrogen producers, eliminate hydrogen consumers, shift the metabolism to favor hydrogen production, increase substrate conversion efficiency, and increase the overall process potential [1, 10, 11, 21]. The organic loading rate (OLR; influent substrate concentration/HRT) is a parameter that evaluates the simultaneous effects of influent substrate concentrations and HRTs when synthetic or real wastewaters are used to produce H_2 in anaerobic reactors [13–18, 22–26]. Previous studies in our research group observed hydrogen production with glucose concentrations of 2000 mg L^{-1} [27–29], 4000 mg L^{-1} [6, 30] and 5000 mg L^{-1} [31]. Increasing glucose concentration to 10 g L^{-1} and 25 g L^{-1} can determine the range where hydrogen-producing acidogenesis shifts to solventogenesis. Therefore, the present study examines the effect of both OLR and alkalinity supplementation on H_2 production in AFBRs with influent glucose concentrations of 10 g L^{-1} (OLRs of 30–240 kg COD m^{-3} day^{-1}) and 25 g L^{-1} (OLRs of 75–600 kg COD m^{-3} day^{-1}).

2. Materials and methods

2.1. Anaerobic fluidized bed reactors and feed composition

A schematic diagram of the two identical jacketed AFBRs used in this study is presented in **Figure 1**. The reactors were constructed with a transparent acrylic tube, within 5.3 cm of

http://dx.doi.org/10.5772/67101

internal diameter and 190 cm of height, and filled with expanded clay (diameter = 2.8–3.3 mm, density = 1.5 g cm^{-3}). Each AFBR was equipped with a water jacket that recirculated heated water from a thermostatic bath to maintain the temperature at 30°C. The AFBRs were fed with synthetic wastewater containing glucose (10 and 25 g L^{-1}) as the main carbon source supplemented with the following nutrients: SeO_2, 0.07 mg L^{-1}; $CoCl_2 \cdot 2H_2O$, 0.08 mg L^{-1}; $FeCl_3 \cdot 6H_2O$, 0.5 mg L^{-1}; $NiSO_4 \cdot 6H_2O$, 1 mg L^{-1}; $FeSO_4 \cdot 7H_2O$, 5 mg L^{-1}; K_2HPO_4, 21.7 mg L^{-1}; $Na_2HPO_4 \cdot 2H_2O$, 33.4 mg L^{-1}: $CaCl_2 \cdot 6H_2O$, 47 mg L^{-1}; KH_2PO_4, 85 mg L^{-1}; and $CO(NH_2)_2N_2O$, 125 mg L^{-1}. In order to control the pH of the reactors at 5.0–5.5, hydrochloric acid (10 M) and sodium bicarbonate (1 g L^{-1}) were also used [6].

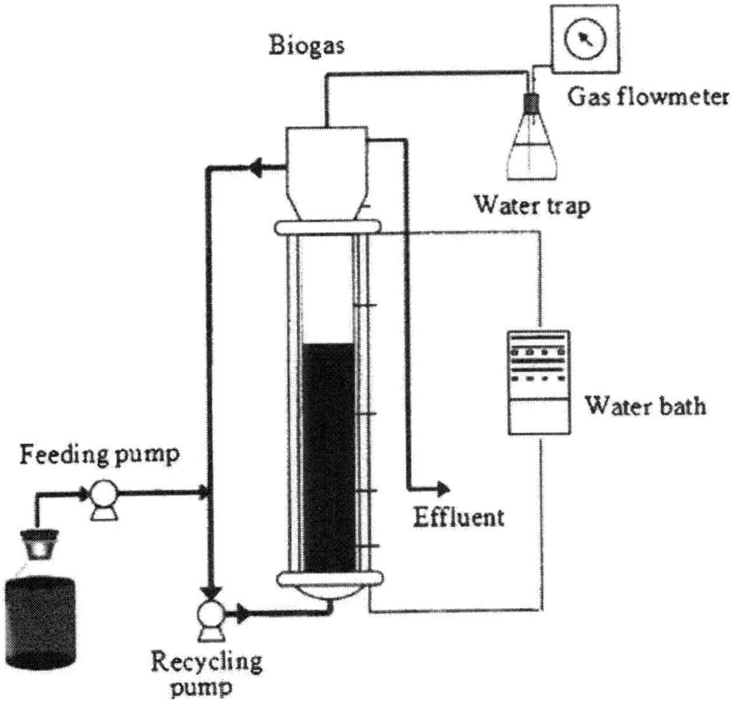

Figure 1. Schematic description of the AFBR.

2.2. Heat treatment of inoculum, AFBR setup and operation conditions

The AFBRs were inoculated with sludge from an upflow anaerobic sludge blanket (UASB) reactor treating swine wastewater effluent. The sludge was heat treated for 10 min at 90°C according to the methodology of Kim et al. [25] in order to eliminate hydrogen consumers and select for endospore producers. The reactors were inoculated at a rate of 10% of the sludge feed volume.

The total liquid flow rate into the AFBRs was maintained at 128 L h^{-1} (expansion = 30%). This flow rate produced a superficial velocity 1.30 times greater than the minimum fluidization velocity. At first, in order to activate the H_2-producing biomass, the two AFBRs were operated in batch mode for 48 h while periodically recording the substrate consumption by microorganisms. When the activation period was over, the reactors were operated in continuous mode with an HRT of 8 h, which was then decreased stepwise to 6 h, 4 h, 2 h, and 1 h. The composition of the gaseous products (H_2 and CO_2) and soluble metabolites (volatile organic acids and alcohols) produced during fermentative H_2 production was monitored as a function of time.

To facilitate discussion of the results and to identify the reactors, each reactor was named according to the influent glucose concentration: the reactor operated with 10 g L^{-1} glucose was named "R10," and the reactor operated with 25 g L^{-1} glucose was named "R25."

2.3. Chemical analyses

The GOD-PAP enzymatic method [32] was used to determine the glucose concentrations. Total solids (TS), volatile suspended solids (VSS), total volatile solids (TVS), and chemical oxygen demand (COD) analyses were performed according to Standard Methods for the Examination of Water and Wastewater [33].

A gas chromatograph (GC-2010, Shimadzu, Tokyo, Japan) equipped with a thermal conductivity detector (TCD) was used to determine the biogas composition. Argon was used as the carrier gas with a Carboxen 1010 PLOT column (30 m long × 0.53 mm internal diameter). A gas chromatograph (GC-2010, Shimadzu, Tokyo, Japan) equipped with a flame ionization detector (FID) was used to determine volatile organic acids and alcohols. The GC used a COMBI-PAL headspace sample introduction system (AOC 5000 model) and HP-INNOWAX column (30 m long × 0.25 mm internal diameter × 0.25 mm film thickness) [32].

A gas meter (type TG1; Ritter Inc., Germany) was used to measure the amount of H_2 generated.

3. Results and discussion

3.1. Effect of OLR on H_2 production

Figure 2 presents the variation in pH effluent as a function of OLR for the two AFBRs used in this study. The pH remained stable throughout the system operation within the operating range of acidogenic anaerobic systems, i.e., between 3.7 in Barros et al. [6], 3.4 and 3.6 in R10, and 3.3 and 3.5 in R25. The influent pH remained between 5.2 and 5.9 in Barros et al. [6], 4.8 and 5.6 in R10, and 5.5 and 5.9 in R25 (**Figure 2**).

Figure 3 presents the variation in glucose conversion as a function of OLR for the AFBRs used in this study. To estimate glucose consumption during fermentation, glucose levels were measured in the fermentation medium (**Figure 3**). Glucose consumption by microorganisms was recorded

http://dx.doi.org/10.5772/67101

at all OLR intervals in both AFBRs. The data indicate that glucose conversion decreased with the increase of OLR at all concentrations. For reactor R10, when OLR was increased from 30–120 kg COD m^{-3} day^{-1}, glucose conversion decreased from 57 to 36%, but when OLR increased to 240 kg COD m^{-3} day^{-1}, glucose conversion increased to 41%. For reactor R25, when OLR increased from 75 to 600 kg COD m^{-3} day^{-1}, glucose conversion decreased from 36 to 20%.

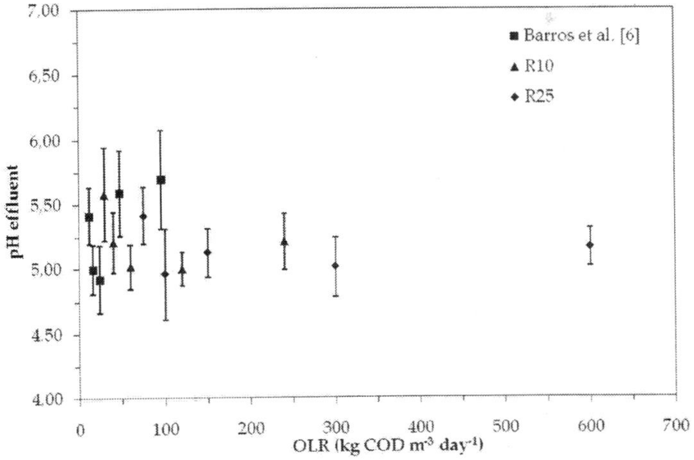

Figure 2. pH effluent as a function of the OLR for the AFBRs.

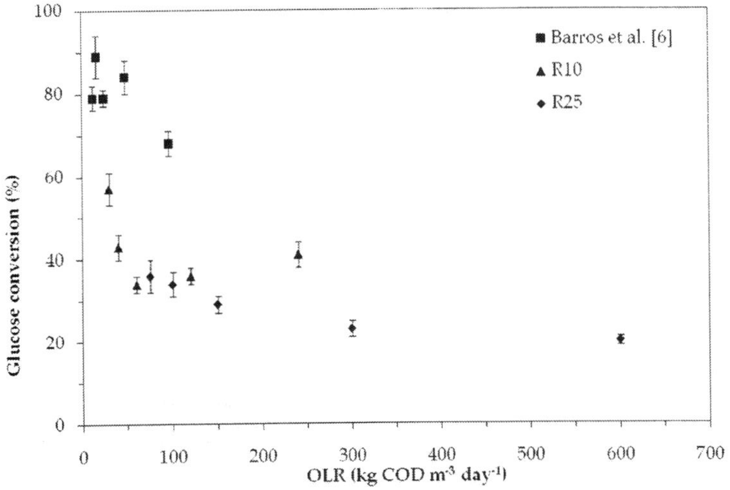

Figure 3. Glucose conversion as a function of the OLR for the AFBRs.

Figure 4 presents the variation in the hydrogen production rate (HPR) as a function of OLR for the two AFBRs used in this study. Similar to the results of Barros et al. [6] for an AFBR with expanded clay as the support material, an influent glucose concentration of 4 g L^{-1}, and alkalinity supplementation (values presented in **Figure 2**), the HPR values for R10 increased linearly from 0.12 to 1.58 L h^{-1} L^{-1} when OLR increased from 30 to 240 kg COD m^{-3}. By contrast, a linear relationship between HPR and OLR was not observed in R25 for OLR ranging from 75 to 600 kg COD m^{-3}. The maximum HPR values were 1.58 and 0.84 L h^{-1} L^{-1} for reactors R10 and R25, respectively.

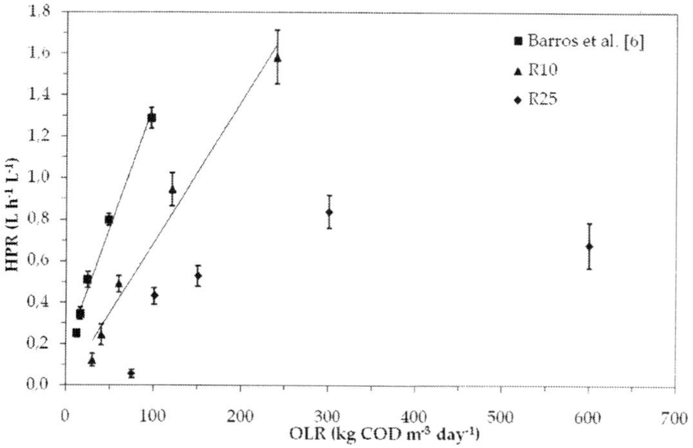

Figure 4. HPR as a function of the OLR for the AFBRs.

Figure 5 presents the variation in HY as a function of OLR for the two AFBRs used in this study. The HY values increased with increasing OLR in both reactors. For reactor R10, when OLR was increased from 30 to 120 kg COD m^{-3} day^{-1}, HY increased significantly from 0.48 to 1.32 mol H_2 mol^{-1} glucose, but when OLR increased to 240 kg COD m^{-3} day^{-1}, HY decreased to 1.04 mol H_2 mol^{-1} glucose. For reactor R25, when OLR increased from 75 to 300 kg COD m^{-3} day^{-1}, the increase in HY was less significant, i.e., from 0.44 to 0.63 mol H_2 mol^{-1} glucose, but when OLR increased to 600 kg COD m^{-3} day^{-1}, the yield decreased to 0.56 mol H_2 mol^{-1} glucose.

Figure 6 presents the variation in H_2 content as a function of OLR for the two AFBRs used in this study. In reactors R10 and R25, the behavior of the H_2 content also varied according to changes in OLR. The hydrogen content of the biogas increased with increasing OLR in both reactors, with a higher H_2 content for HRT 1 h (240 and 600 kg COD m^{-3} day^{-1}, respectively). The H_2 content ranged from 8 to 58% for R10 and 10 to 57% for R25.

The glucose conversion, HPR, HY, and H_2 content of the reactors are consistent with the results of several studies conducted using AFBRs [6, 18, 27, 28, 30–32, 34, 35].

http://dx.doi.org/10.5772/67101

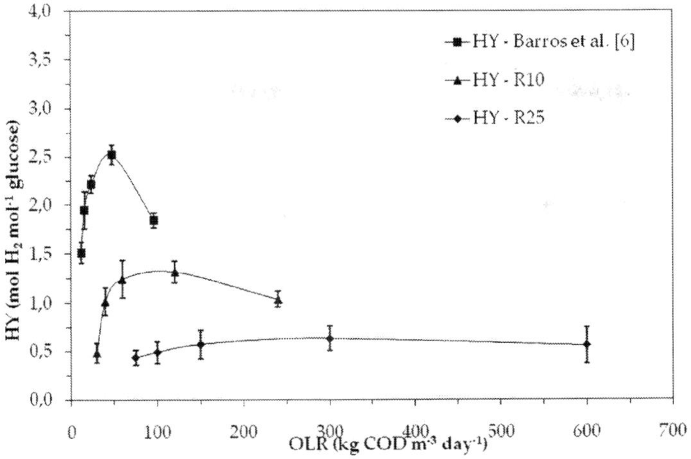

Figure 5. HY as a function of the OLR for the AFBRs.

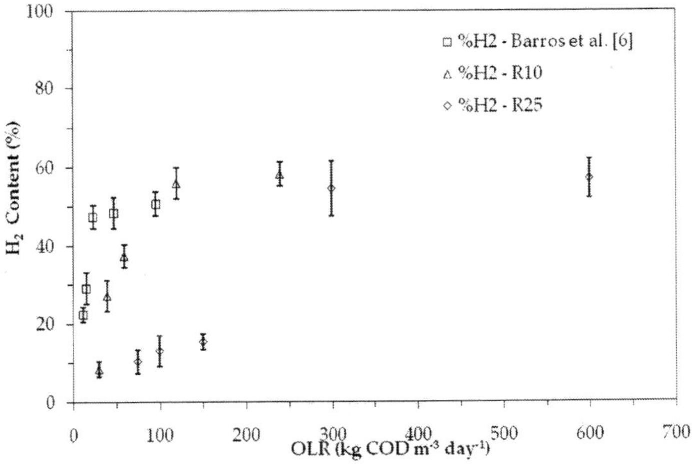

Figure 6. H_2 content as a function of the OLR for the AFBRs.

Table 1 compares studies that evaluated OLR and HY. Studies that observed a decrease in HY with increasing OLR used an OLR range of 6–833.3 kg COD m^{-3} day^{-1} and reported HYs of 4.26–0.81 mol H_2.mol^{-1} substrate. By contrast, studies that observed an increase in HY with increasing OLR worked with an OLR range of 13.5–480 kg COD m^{-3} day^{-1} and reported HYs of 0.94–2.49 mol H_2 mol^{-1} substrate.

Study	Substrate	OLR (kg m^{-3} d^{-1})		HY (mol H_2 mol^{-1} substrate)	
		Low	High	Low OLR	High OLR
Lower OLR improves H_2 production					
Yu et al. [36]	Rice winery	168	432	1.89	1.79
Van Ginkel and Logan [24]	Glucose	25.6	76.8	2.20	2.00
Van Ginkel and Logan [37]	Glucose	6	24	2.80	2.20
Kyazze et al. [15]	Sucrose	22.4	112.2	1.65	0.81
Lin et al. [38]	Sucrose	34.7	833.3	4.26	2.31
Davila-Vasquez et al. [39]	Cheese whey	54	138.6	2.4	1.0
Higher OLR improves H_2 production					
Lin et al. [18]	Sucrose	13.5	107.9	1.69	2.49
	Sucrose	20	160	1.34	2.17
Zhang et al. [35]	Glucose	60	480	0.94	1.19
Shida et al. [27]	Glucose	6	48	1.84	2.29
Perna et al. [17]	Cheese whey	22	37	0.5	0.67

Adapted from Kraemer and Bagley [26].

Table 1. Comparison of the studies that varied the OLR by changing the substrate concentration.

According to Kraemer and Bagley [26], the reason for the variations of H_2 yield at lower or higher OLRs is unknown. High OLR values may reduce the production of H_2 by (1) increasing inhibition by volatile fatty acids (VFAs) with increasing OLR, (2) decreasing thermodynamic regulation due to lower dissolved H_2 concentrations at lower OLRs, (3) affecting acetogenic activity, and (4) increase CO_2 inhibition by increasing the concentration of dissolved CO_2.

Inhibition by VFAs at high OLR values appears to be a valid explanation. The ability of added external VFA to reduce or inhibit the production of H_2 in mixed-culture and continuous-flow systems has been studied, and there is consensus that butyrate increases higher inhibition than the acetate [18, 24, 40].

H_2 production was also assessed with or without the addition of sodium bicarbonate as an alkalizing agent. The effect of the alkalizing agent on pH was important for controlling the hydrogen content and CO_2 in the system. The high HY in the absence of a buffering agent can be attributed to the pH range of the reactor and the CO_2 concentrations produced at steady bicarbonate concentrations [41–44].

3.2. Soluble microbial products

Table 2 presents the distribution of soluble microbial products (SMPs) with increasing glucose concentration and increasing OLRs in the AFBRs. The molar fractions of acetic and butyric acid were the largest by percentage. Barros et al. [6] for an AFBR with expanded clay as the support material, an influent glucose concentration of 4 g L^{-1}, and alkalinity supplementation (values presented in **Table 2**) observed a descending order of products of acetate (32.99–46.81%), butyrate (37.30–41.49%), ethanol (10.18–22.95%), and propionate (1.26–4.90%). In our reactor R10, the products in descending order were ethanol (45.54–71.54%), acetate (27.11–50.63%), butyrate (2.91–31.03%) and methanol (0.00–14.41%). In reactor R25, the products in descending order were ethanol (48.00–71.54%), acetate (12.05–37.43%), butyrate (01.02–29.09%), and methanol (0.00–14.41%) (**Table 2**).

Reactor	OLR (kg COD m^{-3} day^{-1})	HAc (mM)	HBu (mM)	HPr (mM)	EtOH (mM)	MetOH (mM)	TVFA (mM)	TSolv (mM)	HAc/HBu
Barros et al. [6]	12	6.25	7.67	0.68	4.35	0	14.60	4.35	0.81
	16	10.00	11.08	0.34	5.43	0	21.42	5.43	0.90
	24	12.50	11.08	0.41	2.72	0	23.98	2.72	1.13
	48	12.83	10.63	0.68	4.35	0	24.13	4.35	1.21
	96	9.06	8.35	1.01	2.28	0	18.42	2.28	1.08
R10	30	10.73	0.62	0.00	9.35	0.49	11.34	9.84	17.42
	40	7.23	1.57	0.00	10.62	1.44	8.80	12.06	4.62
	60	9.66	3.53	0.00	12.70	9.58	13.20	22.28	2.74
	120	6.37	5.75	0.00	10.70	0.00	12.11	10.7	1.11
	240	6.65	7.61	0.00	10.27	0.00	14.27	10.27	0.87
R25	75	9.04	2.60	0.13	11.59	0.78	11.77	12.37	3.47
	100	17.39	2.70	1.20	21.24	4.10	21.30	25.34	6.43
	150	6.64	1.11	0.00	39.42	7.94	11.70	47.36	6.01
	300	5.92	3.53	0.00	10.65	2.01	9.45	12.66	1.68
	600	4.88	6.18	0.00	10.18	0.00	11.06	10.18	0.79

HAc acetate, HBu butyrate, HPr propionate, EtOH ethanol, MetOH methanol, TVFA total volatile fatty acids, TVFA HAc + HBu + HPr, SMP TVFA + EtOH + MetOH, HAc/SMP molar acetate-to-SMP ratio, HBu/SMP molar butyrate-to-SMP ratio, HPr/SMP molar propionate-to-SMP ratio, EtOH/SMP molar ethanol-to-SMP ratio, MetOH/SMP molar methanol-to-SMP ratio, HAc/HBu molar acetate-to-butyrate ratio

Table 2. Effect of glucose concentration and OLR on the SMP distribution in the AFBRs.

Previous studies employing conditions similar to those used in the present study observed the production of similar metabolites, although differences in the distributions of the metabolites were observed [6, 18, 27, 28, 30–32, 34, 35].

The reactors R10 and R25 produced higher amounts of solvents, such as MetOH and EtOH in the R25 reactor. The higher EtOH concentrations observed in R10 and R25 are similar to the results of Wu et al. [34]. However, our recent studies [6, 27, 29] that used the same medium composition, inoculum, and support material have significantly different results. Barros et al. [6] with an influent glucose concentration of 4 g L^{-1}, and alkalinity supplementation, observed ethanol percentages lower than 22.95% at the beginning of the operation and subsequently decreased and stabilized to 11%. EtOH production is considered unfavorable for hydrogen metabolite production because no H_2 is consumed or produced (Eq. (1)):

$$C_6H_{12}O_6 \rightarrow 2\,CH_3CH_2OH + 2\,CO_2 \tag{1}$$

Propionate was only detected during the operation of the reactor containing 25 g L^{-1}, with maximum concentration of 1.20 mM in the OLR of 100 kg COD m^{-3} day^{-1}. Propionic acid production was not observed in AFBRs with influent glucose concentration of 2 g L^{-1} [27, 29]. Zhang et al. [35] suggested that the absence of propionic acid may be due to inhibition of the activity of the bacteria that form this acid under low pH conditions; these bacteria may be sensitive to both low HRTs and high OLRs. Moreover, the absence of propionic acid production ensures greater production of hydrogen due to the lower consumption of H_2 for forming propionate (Eq. (2)):

$$C_6H_{12}O_6 + 2\,H_2 \rightarrow CH_3CH_2COOH + 2\,H_2O \tag{2}$$

Both HAc and HBu are soluble metabolites favoring H_2 production because these products are generated during H_2 production (Eqs. (3) and (4)):

$$C_6H_{12}O_6 + 2\,H_2O \rightarrow 2\,CH_3COOH + 2\,CO_2 + 4\,H_2 \tag{3}$$

$$C_6H_{12}O_6 \rightarrow CH_3CH_2CH_2COOH + 2\,CO_2 + 2\,H_2 \tag{4}$$

Previous studies have observed that H_2 production increases with the molar ratio of HAc/HBu [45, 46]. **Table 2** presents the variation of the HAc/HBu ratio in R10 and R25. Barros et al. [6] for an influent glucose concentration of 4 g L^{-1}, and alkalinity supplementation, observed the best proportion of soluble metabolites and therefore a higher yield of hydrogen, with molar ratios of HAc/HBu ranging from 0.81 to 1.21 for OLRs varied 12–96 kg COD m^{-3} day^{-1}, respectively, but decreasing to 1.08 for an OLR of 96 kg COD m^{-3} day^{-1}. In our R25, similar behavior of Barros et al. [6] were obtained, but in R10 HAc/HBu ratio decreased from 17.42 to 0.87 when the OLRs increased from 30 to 240 kg COD m^{-3} day^{-1}.

According to Hafez et al. [45], when OLR increased from 6.5 to 103 g COD L^{-1} day^{-1}, acetate and butyrate were the main liquid products, with trace concentrations of ethanol and no detectable lactate, whereas in the OLR range of 154–206 g COD L^{-1} day^{-1}, the concentrations of propionate, isovalerate, valerate, and ethanol increased markedly. The steady-state average molar ratios of acetate/butyrate were 2.3, 2.3, 2.0, and 2.2 for OLRs of 6.5, 25.7, 51.4, and 103 g COD L^{-1} day^{-1}, respectively, but decreased to 1.1 for OLRs of 154 and 206 g COD L^{-1} day^{-1}.

According to Prakasham et al. [47], at lower substrate conditions with the limitation of substrate concentration, increasing glucose concentration progressively increases H_2 production because of effective metabolism and further H_2 production process. However, higher concentrations can also negatively impact H_2 production. When the H_2 yield observed value reduced

http://dx.doi.org/10.5772/67101

because the glucose concentration was above the optimum value, a limited glucose utilization occurred, or a shift in the metabolic pathway from the acidogenic phase to a solventogenic phase took place.

Hydrogen and CO_2 were the only gaseous metabolites during all stages of the experiment. NO CH_4 was detected in the biogas from either reactor. The combination of heat treatment of the inoculum and operation under acidogenic pH conditions inhibited the methanogenic activity responsible for the consumption of hydrogen in the system. Furthermore, the results in the literature suggest that manipulating some operational parameters such as the HRT contributes to the elimination of methanogenic archaea in the reactors.

According to Chen et al. [48], these microorganisms fail to thrive in part because the maximum specific growth rate of methanogenic archaea ($\mu_{maximum} = 0.0167\ h^{-1}$) is significantly lower than that of acidogenic microorganisms ($\mu_{maximum} = 0.083\ h^{-1}$). Thus, methanogenic microorganisms are unable to reproduce or remain in equilibrium under these conditions, resulting in their removal from the reactor.

3.3. COD removal and carbon balance

The carbon balance in the reactors can be calculated by Eq. (5) according to Gavala et al. [49]. The comparison between measured and calculated COD concentrations for each steady state is also presented. The COD calculations were performed as the following: the products ($COD_{products}$) and the glucose ($COD_{glucose}$) COD concentrations were calculated according to Eqs. (5) and (6), respectively. The $COD_{residual}$ was calculated after subtraction of the sum of the $COD_{products}$ and $COD_{glucose}$ from the $COD_{measured}$ (Eq. (3)).The COD_{others} corresponds to the non-identified metabolic products during glucose fermentation:

$$COD_{products} = a \,.\left(\frac{mmolHAc}{1}\right).\,64\frac{mgCOD}{mmolHAc} + b \,.\left(\frac{mmolHBu}{1}\right).\,160\frac{mgCOD}{mmolHBu}\left(\frac{mmolHAc}{1}\right)$$
$$+ c.\left(\frac{mmolHPr}{1}\right).\,112\frac{mgCOD}{mmolPr} + d \,.\left(\frac{mmolMetOH}{1}\right).\,48\frac{mgCOD}{mmolMetOH}$$
$$+ e \,.\left(\frac{mmolEtOH}{1}\right).\,96\frac{mgCOD}{mmolEtOH} \tag{5}$$

where a, b, c, d, and e are the measured concentrations of the acetic acid, butyric acid, propionic acid, methanol, and ethanol, respectively.

$$COD_{glucose} = f \,.\left(\frac{mg\ Glucose}{1}\right)\frac{192\ mg\ COD}{180\ mg} \tag{6}$$

where f is the measured concentration of glucose.

The difference between $COD_{measured}$ and COD based on SMP may be attributed to the presence of other soluble metabolites that were not detected, e.g., lactic acid and formic acid, because the chromatographic method of headspace extraction used in this study only detects alcohols and volatile acids.

This difference was calculated based on Eq. (7):

$$COD_{others} = COD_{measured} - \left(COD_{products} + COD_{glucose}\right) \tag{7}$$

Table 3 presents influent and effluent COD values and standard deviations as well as efficiencies for all reactors. Influent COD represents glucose added to the wastewater and carbonaceous matter present in urea. Effluent COD corresponds to the carbonaceous matter in the effluent that was oxidized. Carbonaceous matter present in the effluent consists of non-consumed glucose; soluble metabolites, e.g., organic acids, solvents, and other intermediary compounds; and biomass detached from the support medium.

	OLR (kg COD m^{-3} day^{-1})	Influent COD (mg L^{-1})	Effluent COD (mg L^{-1})	COD removal (%)
Barros et al. [6]	12	4216 ± 210	3788 ± 153	10 ± 6
	16	4140 ± 206	3349 ± 146	19 ± 9
	24	4139 ± 270	3718 ± 165	10 ± 4
	48	4487 ± 220	3805 ± 191	15 ± 2
	96	4312 ± 226	3680 ± 136	15 ± 4
R10	30	11,298 ± 954	8617 ± 457	24 ± 5
	40	10,439 ± 843	9056 ± 419	13 ± 6
	60	10,693 ± 977	8639 ± 433	19 ± 3
	120	10,175 ± 799	8589 ± 447	16 ± 2
	240	10,969 ± 901	8705 ± 512	21 ± 2
R25	75	26,126 ± 1024	20,202 ± 978	23 ± 3
	100	26,447 ± 1201	22,352 ± 883	15 ± 2
	150	27,285 ± 1392	22,207 ± 791	19 ± 2
	300	26,116 ± 1273	23,502 ± 943	10 ± 1
	600	28,216 ± 1321	25,242 ± 967	11 ± 2

Table 3. Influent COD, effluent COD, and COD removal in AFBRs.

The theoretical effluent COD was calculated based on stoichiometric relationships for oxidation of glucose, acetic acid, butyric acid, propionic acid, biomass, ethanol, and methanol to estimate the carbon balance. Theoretical COD values for the remaining glucose, soluble metabolites, and biomass as well as the difference between the theoretical total COD and the COD measured for all reactors are presented in **Table 4**.

In the reactor operated by Barros et al. [6], this difference varied between 12 and 350 mg L^{-1}, which corresponded to a variation of 0.34 and 9.19%. The reactor R10 showed a difference ranging from 91 to 301 mg L^{-1} (variation of 1.05 and 3.28%), whereas in the reactor R25, the difference varied between 17 and 1026 mg L^{-1} (variation of 0.07 and 4.62%).Those differences may be accredited to the presence of other metabolites such as lactic acid and formic acid that were not detected, probably due to the chromatographic method performed (headspace extraction), considering that this method can only detect volatile acids and alcohols.

Reactor	OLR (kg COD m^{-3} day^{-1})	HRT (h)	$COD_{t,\ glucose}$ (mg L^{-1})	$COD_{t,\ acetate}$ (mg L^{-1})	$COD_{t,\ butyrate}$ (mg L^{-1})	$COD_{t\ propionate}$ (mg L^{-1})	$COD_{t,\ biomass}$ (mg L^{-1})	$COD_{t,\ ethanol}$ (mg L^{-1})	$COD_{t,\ methanol}$ (mg L^{-1})	$COD_{t\ total}$ (mg L^{-1})	$COD_{measured}$ (mg L^{-1})	COD_{others} (mg L^{-1})
Barros et al. [6]	12	8	946	245	1382	0	192	90	24	3405	3788	39
	16	6	475	192	1000	0	157	203	105	3157	3349	32
	24	4	901	320	1563	0	161	215	0	3432	3719	12
	48	2	666	320	1763	0	155	629	0	3455	3805	350
	96	1	1394	235	964	0	181	573	0	3556	3680	124
R10	30	8	4514	757	645	0	148	1540	940	8545	8617	159
	40	6	5807	438	705	0	157	457	564	8129	9056	104
	60	4	6935	291	551	0	140	631	0	8548	8639	91
	120	2	6659	364	858	0	134	585	0	8600	8589	254
	240	1	6639	294	699	0	168	959	104	8862	8705	301
R25	75	8	17,177	1210	271	47	148	2178	144	21,174	20,202	1026
	100	6	16,590	769	330	0	145	4825	760	23,419	22,352	486
	150	4	19,454	452	425	0	141	1692	275	22,439	22,207	107
	300	2	21,122	373	636	0	134	1360	96	23,722	23,502	17
	600	1	22,996	269	751	0	168	1023	0	25,206	25,242	35

Table 4. Theoretical COD values of soluble metabolites, biomass COD, and effluent COD measured in AFBRs.

The largest variation between COD measured in the effluent and the theoretical COD (corresponding to glucose, soluble metabolites, and biomass in the effluent) was 9.19% based on the results obtained from the carbon balance. However, according to Standard Methods [33], the determination of metabolites and COD produces errors of close to 10%. For that reason, this variation may be attributed to the margin of error of the determination methods used.

4. Conclusions

Satisfactory performance for H_2 production was observed in the anaerobic fluidized bed reactor containing 10 g L^{-1} glucose. However, in the reactor containing 25 g L^{-1} glucose, the yield was limited.

The HPR had a linear increase with OLR, with the exception of reactor operated with 25 g L^{-1} glucose. The maximum HPR was 1.58 L h^{-1} L^{-1} obtained in the reactor with 10 g L^{-1} glucose for

OLR of 240 kg COD m^{-3} day^{-1} (HRT = 1 h). The maximum HY was 1.32 mol H_2 mol^{-1} glucose obtained in the reactor with 10 g L^{-1} glucose for HRT 2 h (OLR = 240 kg COD m^{-3} day^{-1}).

The H_2 production with addition of sodium bicarbonate was important to control the pH and CO_2 system. The reactors operated at high glucose concentrations (10 and 25 g L^{-1}) showed higher proportions of solvents.

Author details

Eduardo Lucena Cavalcante de Amorim[1,2], Leandro Takano Sader[3], Lucas Rodrigues Ramos[3] and Edson Luiz Silva[3]*

*Address all correspondence to: edsilva@ufscar.br

1 Technology Center, Federal University of Alagoas, Maceió, AL, Brazil

2 Department of Hydraulics and Sanitation, University of São Paulo, São Carlos, SP, Brazil

3 Department of Chemical Engineering, Federal University of São Carlos, São Carlos, SP, Brazil

References

[1] Bartacek J, Zabranska J, Lens PNL. Developments and constraints in fermentative hydrogen production. Biofuels, Bioproducts Biorefining. 2007;**1**:201–214. DOI: 10.1002/bbb.

[2] Mohan SV. Harnessing of biohydrogen from wastewater treatment using mixed fermentative consortia: Process evaluation towards optimization. International Journal of Hydrogen Energy. 2009;**34**:7460–7474. DOI: 10.1016/j.ijhydene.2009.05.062.

[3] Reddy MV, Amulya K, Rohit MV, Sarma PN, Mohan SV. Valorization of fatty acid waste for bioplastics production using Bacillus tequilensis: Integration with dark-fermentative hydrogen production process. International Journal of Hydrogen Energy. 2014;**39**:7616–7626. DOI: 10.1016/j.ijhydene.2013.09.157.

[4] Sarma SJ, Brar SK, Bihan YL, Buelna G. Liquid waste from bio-hydrogen production—a commercially attractive alternative for phosphate solubilizing bio-fertilizer. International Journal of Hydrogen Energy. 2013;**38**:8704–8707. DOI: 10.1016/j.ijhydene.2013.05.032

[5] Sarma SJ, Pachapur V, Brar SK, Bihan YL, Buelna G. Hydrogen biorefinery: Potential utilization of the liquid waste from fermentative hydrogen production. Renewable and Sustainable Energy Reviews. 2015;**50**:942–951. DOI: 10.1016/j.rser.2015.04.191.

[6] Barros AR, Amorim ELC, Reis CM, Shida GM, Silva EL. Biohydrogen production in anaerobic fluidized bed reactors: Effect of support material and hydraulic retention time. International Journal of Hydrogen Energy. 2010;**35**:3379–3388. DOI: 10.1016/j.ijhydene.2010.01.108.

[7] Rosa PRF, Santos SC, Silva EL. Different ratios of carbon sources in the fermentation of cheese whey and glucose as substrates for hydrogen production and ethanol production in continuous reactors. International Journal of Hydrogen Energy. 2014;**39**:1288–1296. DOI: 10.1016/j.ijhydene.2013.11.011.

[8] Rosa PRF, Santos SC, Sakamoto IK, Varesche MBA, Silva EL. The effects of seed sludge and hydraulic retention time on the production of hydrogen from cassava processing wastewater and glucose mixture in an anaerobic fluidized bed reactor. International Journal of Hydrogen Energy. 2014;**39**:13118–13127. DOI: 10.1016/j.ijhydene.2014.06.152.

[9] Santos SC, Rosa PRF, Sakamoto IK, Varesche MBA, Silva EL. Organic loading rate impact on biohydrogen production and microbial communities at anaerobic fluidized thermophilic bed reactors treating sugarcane stillage. Bioresource Technology. 2014;**159**:55–63. DOI: 10.1016/j.biortech.2014.02.051.

[10] Li CL, Fang HHP. Fermentative hydrogen production from wastewater and solid wastes by mixed cultures. Critical Reviews in Environmental Science and Technology. 2007;**37**:1–39. DOI: 10.1080/10643380600729071.

[11] Wang J, Wan E. Factors influencing fermentative hydrogen production: A review. International Journal of Hydrogen Energy. 2009;**34**:799–811. DOI: 10.1016/j.ijhydene.2008.11.015.

[12] Beckers L, Masset J, Hamilton C, Delvigne F, Toye D, Crine M, Thonart P, Hiligsmann S. Investigation of the links between mass transfer conditions, dissolved hydrogen concentration and biohydrogen production by the pure strain *Clostridium butyricum* CWBI1009. Biochemical Engineering Journal. 2015;**98**:18–28. DOI: 10.1016/j.bej.2015.01.008.

[13] Chen CC, Chen HP, Wu JH, Lin CY. Fermentative hydrogen production at high sulfate concentration. International Journal of Hydrogen Energy. 2008;**33**:1573–1578. DOI: 10.1016/j.ijhydene.2007.09.042.

[14] Shen L, Bagley DM, Liss SN. Effect of organic loading rate on fermentative hydrogen production from continuous stirred tank and membrane bioreactors. International Journal of Hydrogen Energy. 2009;**34**:3689–3696. DOI: 10.1016/j.ijhydene.2009.03.006.

[15] Kyazze G, Martinez-Perez N, Dinsdale R, Premier GC, Hawkes FR, Guwy AJ, Hawkes DL. Influence of substrate concentration on the stability and yield of continuous biohydrogen production. Biotechnology and Bioengineering. 2006;**93**:971–979. DOI: 10.1002/bit.20802.

[16] Sreethawong T, Chatsiriwatana S, Rangsunvigit P, Chavadej S. Hydrogen production from cassava wastewater using anaerobic sequencing batch reactor: Effect of operational parameters, COD:N ratio, and organic acid composition. International Journal of Hydrogen Energy. 2010;**35**:4092–4102. DOI: 10.1016/j.ijhydene.2010.02.030.

[17] Perna V, Castelló E, Wenzel J, Zampol C, Fontes Lima DM, Borzacconi L,Varesche MB, Zaiat M, Etchebehere C. Hydrogen production in an upflow anaerobic packed bed reactor

used to treat cheese whey. International Journal of Hydrogen Energy. 2013;38:54–62. DOI: 10.1016/j.ijhydene.2012.10.022.

[18] Lin CN, Wu SY, Chang, JS. Fermentative hydrogen production with a draft tube fluidized bed reactor containing silicone-gel-immobilized anaerobic sludge. International Journal of Hydrogen Energy. 2006;**31**:2200–2210. DOI: 10.1016/j.ijhydene.2006.05.012.

[19] Jung KW, Kim DH, Kim SH, Shin HS. Bioreactor design for continuous dark fermentative hydrogen production. Bioresource Technology. 2011;**102**:8612–8620. DOI: 10.1016/j.biortech.2011.03.056.

[20] Show K, Lee D, Chang J. Bioreactor and process design for biohydrogen production. Bioresource Technology. 2011;**102**:8524–8533. DOI: 10.1016/j.biortech.2011.04.055.

[21] Barca C, Soric A, Ranava D, Giudici-Orticoni MT, Ferrasse JH. Anaerobic biofilm reactors for dark fermentative hydrogen production from wastewater: A review. Bioresource Technology. 2015;**185**:386–398. DOI: 10.1016/j.biortech.2015.02.063.

[22] Mohammadi P, Ibrahim S, Annuar MSM, Ghafari S, Vikineswary S, Zinatizadeh AA. Influences of environmental and operational factors on dark fermentative hydrogen production: A review. Clean—Soil, Air, Water. 2012;**40**:1297–1305. DOI: 10.1002/clen.201100007.

[23] Tawfik A, Salem A. The effect of organic loading rate on bio-hydrogen production from pre-treated rice straw waste via mesophilic up-flow anaerobic reactor. Bioresource Technology. 2012;**107**:186–190. DOI: 10.1016/j.biortech.2011.11.086.

[24] Van Ginkel SW, Logan BE. Inhibition of biohydrogen production by undissociated acetic and butyric acids. Environmental Science and Technology. 2005;**39**:9351–9356. DOI: 10.1021/es0510515.

[25] Kim S, Han S, Shin H. Effect of substrate concentration on hydrogen production and 16S rDNA-based analysis of the microbial community in a continuous fermenter. Process Biochemistry. 2006;**41**:199–2007. DOI: 10.1016/j.procbio.2005.06.013.

[26] Kraemer JT, Bagley DM. Improving the yield from fermentative hydrogen production. Biotechnology Letters. 2007;**29**:685–695. DOI: 10.1007/s10529-006-9299-9.

[27] Shida GM, Barros AR, Reis CM, Amorim ELC, Damianovic MHRZ, Silva EL. Long-term stability of hydrogen and organic acids production in an anaerobic fluidized-bed reactor using heat treated anaerobic sludge inoculum. International Journal of Hydrogen Energy. 2009;**34**:3679–3688. DOI: 10.1016/j.ijhydene.2009.02.076.

[28] Shida GM, Sader LT, Amorim ELC, Sakamoto IK, Maintinguer SI, Saavedra NK, Varesche MBA, Silva EL. Performance and composition of bacterial communities in anaerobic fluidized bed reactors for hydrogen production: Effects of organic loading rate and alkalinity. International Journal of Hydrogen Energy. 2012;**37**:16925–16934. DOI: 10.1016/j.ijhydene.2012.08.140.

[29] Amorim ELC, Barros AR, Damianovic MHRZ, Silva EL. Anaerobic fluidized bed reactor with expanded clay as support for hydrogen production through dark fermentation of glucose. International Journal of Hydrogen Energy. 2009;**34**:783–790. DOI: 10.1016/j.ijhydene.2008.11.007.

[30] Barros AR, Silva EL. Hydrogen and ethanol production in anaerobic fluidized bed reactors: Performance evaluation for three support materials under different operating conditions. Biochemical Engineering Journal. 2012;**61**:59–65. DOI: 10.1016/j.bej.2011.12.002.

[31] Reis CM, Silva EL. Effect of upflow velocity and hydraulic retention time in anaerobic fluidized-bed reactor used for hydrogen production. Chemical Engineering Journal. 2011;**172**:28–36. DOI: 10.1016/j.cej.2011.05.009.

[32] Amorim ELC, Sader LT, Silva EL. Effect of substrate concentration on dark fermentation hydrogen production using an anaerobic fluidized bed reactor. Applied Biochemistry and Biotechnology. 2012;**166**:1248–1263. DOI: 10.1007/s12010-011-9511-9.

[33] Greenberg AE, Clesceri LS, Eaton AD, editors. Standard methods for the examination for water and wastewater. 20th ed. Washington, DC, USA: APHA, WEF, AWWA; 1998. 1496 p.

[34] Wu SY, Lin CN, Chang JS, Lee KS, Lin PJ. Hydrogen production with immobilized sewage sludge in three-phase fluidized-bed bioreactor. Biotechnology Progress. 2003;**19**:828–832. DOI: 10.1021/bp0201354.

[35] Zhang ZP, Tay JH, Show KY, Yan R, Liang DT, Lee DJ, Jiang WJ. Biohydrogen production in a granular activated carbon anaerobic fluidized bed reactor. International Journal of Hydrogen Energy. 2007;**32**:185–191. DOI: 10.1016/j.ijhydene.2006.08.017.

[36] Yu HQ, Zhu ZH, Hu WR, Zhang HS. Hydrogen production from rice winery wastewater in an upflow anaerobic reactor by using mixed anaerobic cultures. International Journal of Hydrogen Energy. 2002;**27**:1359–1365. DOI: 10.1016/s0360-3199(2)00073-3.

[37] Van Ginkel SW, Logan BE. Increased biological hydrogen production with reduced organic loading. Water Research. 2005;**39**:3819–3826. DOI: 10.1016/j.watres.2005.07.021.

[38] Lin CN, Wu SY, Chang JS, Chang JS. Biohydrogen production in a three-phase fluidized bed bioreactor using sewage sludge immobilized by ethylene-vinyl acetate copolymer. Bioresource Technology. 2009;**100**:3298–3301. DOI: 10.1016/j.biortech.2009.02.027.

[39] Davila-Vasquez G, Cota-Navarro CB, Rosales-Colunga LM. Continuous biohydrogen production using cheese whey: Improving the hydrogen production rate. International Journal of Hydrogen Energy. 2009;**34**:4296–4304. DOI: 10.1016/j.ijhydene.2009.02.063.

[40] Gao S, Wang B, Zhu LL, Han W, Chen H, Li YF. Effect of organic loading rate on fermentative hydrogen production in CSTR. Advanced Marerials Research. 2011;**156–157**:732–736. DOI: 10.4028/www.scientific.net/AMR.

[41] Leite JAC, Fernandes BS, Pozzi E, Barboza M, Zaiat M. Application of an anaerobic packed-bed bioreactor for the production of hydrogen and organic acids. International Journal of Hydrogen Energy. 2008;**33**:579–586. DOI: 10.1016/j.ijhydene.2007.10.009.

[42] Valdez-Vazquez I, Poggi-Varaldo HM. Alkalinity and high total solids affecting H_2 production from organic solid waste by anaerobic consortia. International Journal of Hydrogen Energy. 2009;**34**:3639–3646. DOI: 10.1016/j.ijhydene.2009.02.039.

[43] Choi J, Ahn Y. Biohydrogen fermentation from sucrose and piggery waste with high levels of bicarbonate alkalinity. Energies. 2015;**8**:1716–1729. DOI: 10.3390/en8031716.

[44] Silva AJ, Pozzi E, Foresti E, Zaiat M. The influence of the buffering capacity on the production of organic acids and alcohols from wastewater in anaerobic reactor. Applied Biochemistry and Biotechnology. 2015;**175**:2258–2265. DOI: 10.1007/s12010-014-1424-y.

[45] Hafez H, Nakhla G, El Naggar MH, Elbeshbishy E, Baghchehsaraee B. Effect of organic loading rate on a novel hydrogen bioreactor. International Journal of Hydrogen Energy. 2010;**35**:81–92. DOI: 10.1016/j.ijhydene.2009.10.051.

[46] Wang JL, Wan W. The effect of substrate concentration on biohydrogen production by using kinetic models. Science in China Series B-Chemistry. 2008;**51**:1110–1117. DOI: 10.1007/s11426-008-0104-6.

[47] Prakasham RS, Brahmaiah P, Satish T, SambasivaRao KRS. Fermentative biohydrogen production by mixed anaerobic consortia: Impact of glucose to xylose ratio. International Journal of Hydrogen Energy. 2010;**34**:9354–9361. DOI: 10.1016/j/ijhydene.2009.09.104.

[48] Chen CC, Lin CY, Chang JS. Kinetics of hydrogen production with continuous anaerobic cultures utilizing sucrose as the limiting substrate. Applied Microbiology and Biotechnology. 2001;**57**:56–64. DOI: 10.1007/s002530100747.

[49] Gavala HN, Skiadas IV, Ahring BK. Biological hydrogen production in suspended and attached growth anaerobic reactor systems. International Journal of Hydrogen Energy. 2006;**31**:1164–1175. DOI: 10.1016/j/ijhydene.2005.09.009.

Chapter 13

Production of Biogas and Performance Evaluation of Ultrasonic Membrane Anaerobic System (UMAS) for Palm Oil Mill Effluent Treatment (POME)

Abdurahman Hamid Nour and Azhari Hamid Nour

Additional information is available at the end of the chapter

http://dx.doi.org/10.5772/67602

Abstract

This study proposes a new approach for integrated technology of ultrasonic and membrane for a palm oil mill effluent treatment. This study evaluated the performance of the new design of ultrasonic membrane anaerobic system (UMAS) when a palm oil mill effluent (POME) introduces this approach. To fit kinetic study, six steady states were investigated and the results have shown that the mixed liquor volatile suspended solids (MLVSSs) range from 10,400 to 17,350 mg/l while the mixed liquor suspended solids (MLSSs) range from 13,800 to 22,600 mg/l. Three kinetic models of Monod, Contois, and Chen and Hashimoto were used to evaluate the integrated system at organic loading rates ranging from 1 to 15 kg COD/m^3/day. The percentage efficiency of COD removal was from 92.8 to 98.3%, and hydraulic retention time (HRT) was from 500.8 to 8.6 days. The influent COD concentrations of the POME ranged from 70,400 to 90,200 mg/l.The integrated technology of UMAS is a more attractive one as it avoids membrane fouling problems.

Keywords: membrane, ultrasonic, POME, methane, CO_2, UMAS

1. Introduction

The palm oil industry has grown tremendously in the recent years and accounted for the largest percentage of oil and fats production in the world in 2011. Over the last few decades, the palm oil industry has been growing rapidly. Palm oil has risen to become the most produced and consumed vegetable oil in the world, widely used in food, cosmetic and hygienic products due to its affordable price, efficient production and high oxidative stability [1]. Palm oil is the most produced vegetable oil in the world with a global production of almost 60 million tons and a global vegetable oil market share of more than 35% by weight in 2015 as reported by Hansen et al. [2] and MPOB [3]. The industry continues to generate huge revenues for the producing countries. Accordingly, it is not surprising that the oil palm industry is expected to grow further in the coming years as shown in **Figure 1**.

Over the long term, global palm oil demand shows an increasing trend as an expanding global population gives rise to increased consumption of palm-oil based products world consumption of palm oil [5]. [6] Stated that palm oil industries have been significantly contributing towards the economic growth and increase standard of living among the South East Asian countries. Nowadays, the global production and demand for palm oil are increasing rapidly where the plantations are spreading across Asia, Africa and Latin America. The five leading palm oil producing countries are Indonesia, Malaysia, Thailand, Colombia and Nigeria [7] as shown in **Figure 2**.

The development of palm oil industry in Malaysia has turned into a phenomenal in which the area of plantation expanded from year to year. The country is experiencing a robust development in new oil palm plantations and palm oil mills. This commodity plays a significant role in the Malaysia economic growth [8]. Throughout the year, Malaysia is blessed with favorable weather conditions, which are advantageous for palm oil cultivation [9]. Thus, it is not surprising that the highest yields have been obtained from palms grown in this region, which is far from its natural habitat. Besides, the Malaysian palm oil industry has grown to become a very important agriculture-based industry, where the country is today one of the world's leading producer and exporter of palm oil.

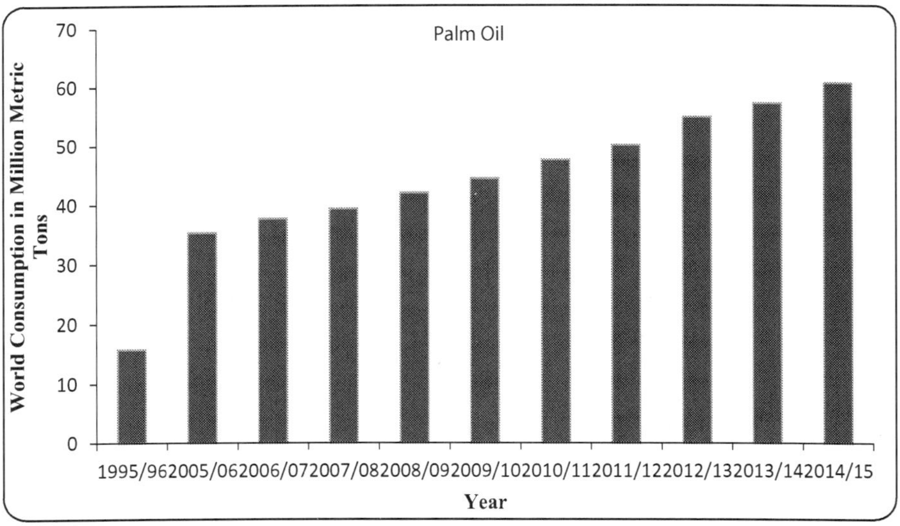

Figure 1. Global consumption of palm oil from 1995/1996 to 2014/2015 (USDA, 2016) [4].

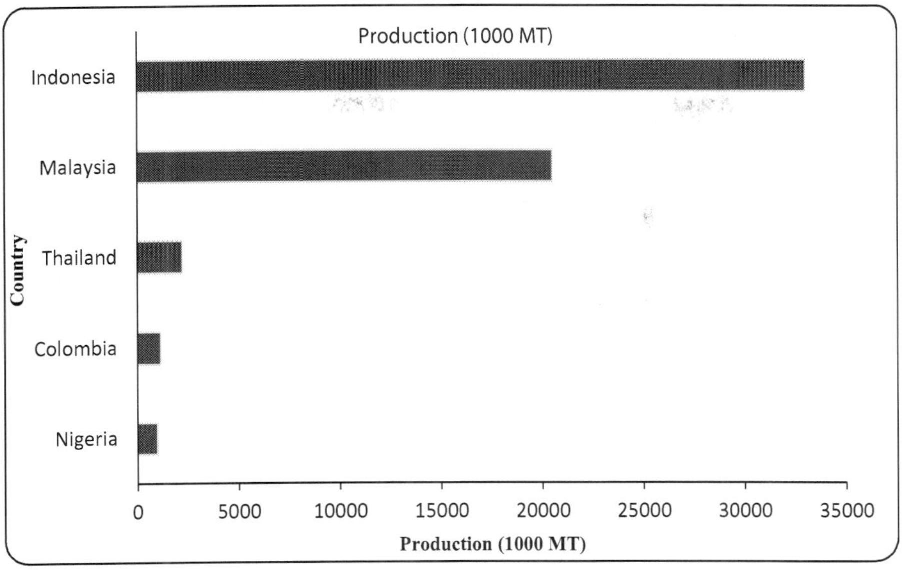

Figure 2. Palm oil production by country [10, 11].

Figure 3 depicts the statistics production of palm oil superseded soybean oil from 13% in 1990 to 28% of total oil and fats production in 2011. This is because oil palm has higher annual oil yield per hectare than other oil seeds crops including soybean [11] and palm oil has a relatively lower price as compared to the major alternative vegetable oils [12]. POME is highly polluted wastewater if not treated properly; it causes a lot of environment issues. POME is a colloidal suspension of 95–96% water, 0.6–0.7% oil and 4–5% total solids including 2–4% suspended solids originating from mixture of a sterilizer condensate, separator sludge and hydrocyclone wastewater [13]. The conventional treatment technology of POME employed in most of the palm oil mills in Malaysia is the ponding system of biological treatment [14–16]. However, coping with the increasing production in most palm oil mills, the undersized biological treatment system is unable to cope with the increased volume of POME [17]. Thus, proper POME treatment is urgently needed to ensure the sustainable economic growth of palm oil industry in Malaysia besides protecting the environment. Several researchers have proposed other biological treatments.

The treatment system includes aerated lagoon system [18], conventional anaerobic digester [19], anaerobic contact process [20], upflow anaerobic sludge blanket (UASB) reactor [17, 19], close tank digester [21], trickling filter, aerobic lagoon system [18], aerobic rotating biological contactor [19] and evaporation process [13].

The main objective of this study was to evaluate the performance and kinetics of the new designed ultrasonic membrane anaerobic system (UMAS) in the treatment of palm oil mill effluent (POME) based on three models [22–24]. **Table 1** shows mathematical expressions for specifics substrate utilization rate for three kinetic models (Monod, Contois, and Chen and Hashimoto).

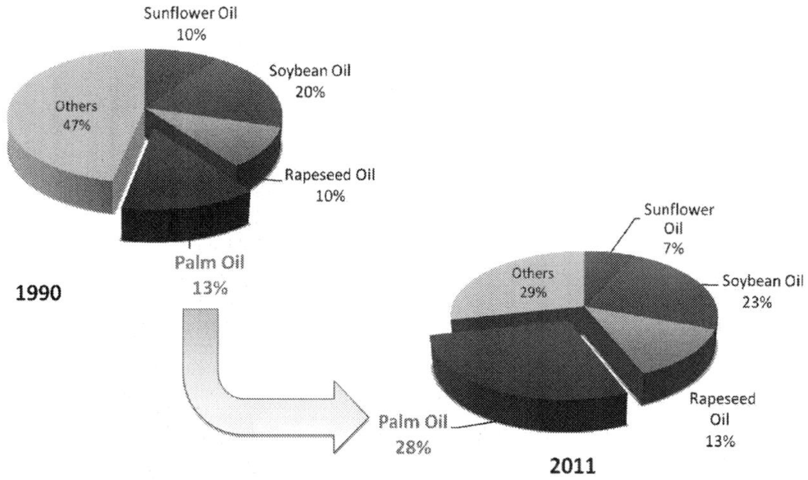

Figure 3. World oil and fat production in 1990 and 2011 [3–5].

Kinetic Model	Equation 1	Equation 2	
Monod	$U = \frac{k S}{k_s + S}$	$\frac{1}{U} = \frac{K_s}{K}(\frac{1}{S}) + \frac{1}{k}$	[22]
Contois	$U = \frac{U_{max} \times S}{Y(B \times X + S)}$	$\frac{1}{U} = \frac{a \times X}{\mu_{max} \times S} + \frac{Y(1+a)}{\mu_{max}}$	[23]
Chen & Hashimoto	$U = \frac{\mu_{max} \times S}{Y K S_o + (1-K) S Y}$	$\frac{1}{U} = \frac{Y K S_o}{\mu_{max} S} + \frac{Y(1-K)}{\mu_{max}}$	[24]

Table 1. Mathematical expressions of specifics substrate utilization rates for known kinetic models.

1.1. Mechanisms of anaerobic digestion

In anaerobic degrading of POME, biogas is formed when microorganisms, especially bacteria, degrade organic material in the absence of oxygen. Biogas consists of 50–75% methane (CH_4), 25–45% carbon dioxide (CO_2) and small amounts of other gases [25–27]. A simplified schematic representation of anaerobic degradation of organic matter is given in **Figure 4**. The AD process can be subdivided into the following four phases, each requires its own characteristic group of microorganisms.

The sequence of reactions involved in the mechanisms of AD is hydrolysis, acidogenesis, acetogenesis and methanogenesis [28]. Hydrolysis is conversion of nonsoluble biopolymers to soluble organic compounds. Acidogenesis is summarized as a conversion of soluble organic compounds to volatile fatty acids (VFA) and CO_2 while acetogenesis is the conversion of VFAs to acetate and H_2 [29]. Methanogenesis represents conversion of acetate and CO_2 plus H_2 to methane and carbon dioxide gas.

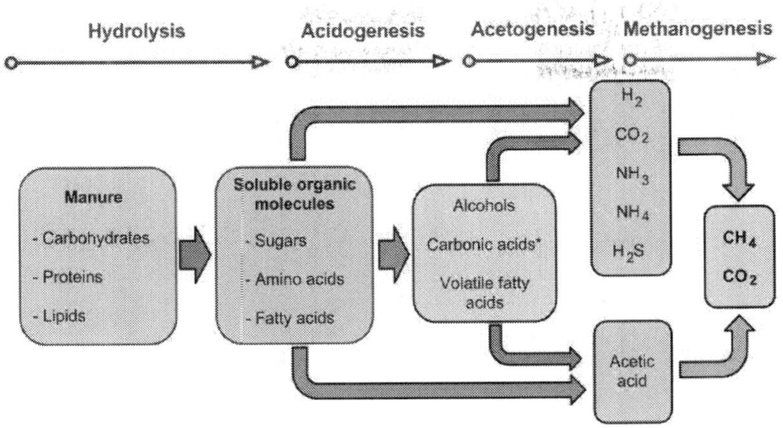

Figure 4. Process stages of anaerobic digestion [30].

2. Materials and methods

2.1. Raw POME wastewater preparation

The raw POME was collected from a near local palm oil mill in Lebah Hillier, Kuantan, Malaysia. The raw POME was stored in a cold room at 4°C before use. Different dilutions of POME were prepared using tap water. The pH of the feed was adjusted to 7.0 using a 6 N NaOH solution.

2.2. UMAS bioreactor operation and experimental setup

A laboratory scale, with an effective 200-L UMAS reactor (**Figure 5**), was used in this study. The UMAS reactor consists of a cross-flow ultrafiltration membrane apparatus, a centrifugal pump and an anaerobic reactor. The total volume of the reactor was 200 L, and the working volume was 150 L. Six multifrequency ultrasonic transducers, operated at 25 KHz, are bonded to two sides of the tank chamber and connected to a Crest Genesis Generator (250 W, 25 KHz; Crest Ultrasonic, Trenton, NJ, USA). The maximum operating pressure on the membrane was 55 bars at 70 WC, and the pH ranged from 2 to 12.

2.3. Analytical methods

The following parameters were analyzed: COD, BOD, pH, VSS and TSS.

Methane gas was determined by gas chromatography with a stainless steel column (200 × 0.3 cm) packed with active carbon (30–60 mesh) using thermal conductivity detection. For TSS, VSS, volatile fatty acids and alkalinity were determined according to the standard methods [31]. The COD was measured using a Hach colorimetric digestion method (Method # 8000, Hach Company, and Loveland, CO, USA). The MLSS and MLVSS were determined by drying the sample at 105 and 550 ± 50°C.

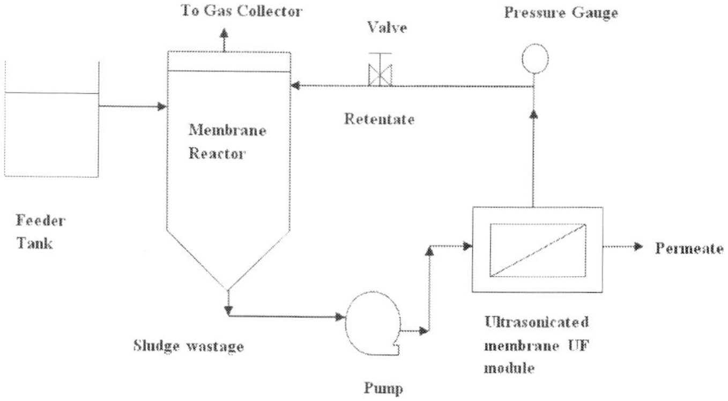

Figure 5. Experimental setup.

2.4. Bioreactor operation

The steady-state performance of ultrasonic membrane anaerobic system (UMAS) was evaluated under different influent COD concentrations (70,400–90,200 mg/l), hydraulic retention time (HRTs) (500.8–14.7 days) and OLR of 1.5–9.0 kg COD/m³/day (**Table 2**). In this study, the system was considered to have achieved steady state when the operating and control parameters were within ±10% of the average value. The produced biogas contained only CO_2 and CH_4, so the addition of sodium hydroxide solution (NaOH) to absorb CO_2 effectively isolated methane gas (CH_4). **Table 2** depicts the results of the application of three known substrate utilization models.

Steady state (SS)	1	2	3	4	5	6
COD feed, mg/L	70,400	73,478	76,200	83,570	86,700	90,200
COD permeate, mg/L	1197	1617	3048	3343	4508	6494
Gas production (L/day)	290	310	340	400	480	540
Total gas yield, L/g COD/day	0.48	0.53	0.58	0.67	0.78	0.81
% Methane	81	78.5	75.6	73.8	68.6	64.6
CH_4 yield, l/g COD/day	0.39	0.54	0.57	0.60	0.64	0.70
MLSS, mg/L	13,800	12,400	13,400	14,800	17,648	22,600
MLVSS, mg/L	10,269	10,751	11,765	13,320	15,530	20,159
% VSS	74.41	86.70	87.80	90.00	88.00	89.20
HRT, day	500.8	60.6	22.6	14.7	11.20	8.6
SRT, day	300	250	180	30.5	20.30	15.80
OLR, kg COD/m³/day	1.0	3.5	6.0	8.5	11.0	15
SSUR, kg COD/kg VSS/day	0.164	0.195	0.252	0.263	0.294	0.314
SUR, kg COD/m³/day	0.023	0.724	2.225	4.576	5.685	7.347
Percent COD removal (UMAS)	**98.3**	**97.8**	**96**	**96.0**	**94.8**	**92.8**

Table 2. Summary of results (SS: steady state).

3. Results and discussion

3.1. The performance of ultrasonic membrane anaerobic system (UMAS)

The operating conditions for the ultrasonic membrane anaerobic system (UMAS) over the 500-day experimental setup are given in **Table 2**. The performance evaluation of the integrated ultrasonic membrane anaerobic system (UMAS) was generated at different influent COD concentrations and hydraulic retention times (HRTs). **Table 3** depicted and summarized the kinetic coefficients. For the system results at influent COD concentrations from 70,400 to 90,200 mg/l and pH (6.7–7.8), UMAS was performed well. The mixed liquor volatile suspended solids (MLVSSs) for the first steady state were 10,400 mg/l, whereas the mixed liquor suspended solids (MLSSs) were 13,800 mg/l, equivalent to 75.36% of the MLSS. This low result can be explained due the palm oil mill effluent wastewater contains very high suspended solids.

The volatile suspended solid (VSS) fraction in the reactor at sixth steady state was increased to 89.20%. Results have shown that the long solid retention time (SRT) of UMAS facilitated the decomposition of the suspended solids and their subsequent conversion to methane (CH_4); these findings found by Nagano et al. [32] and Abdurahman et al. [33]. At organic loading rate, OLR of 15 kg COD/m^3/day, the system registered the highest influent of COD 90,200 mg/l at this stage; the UMAS achieved 92.8% COD removal. **Figures 6–8** shown that UMAS can be applied and treat POME efficiently. Among the three models applied, the Monod and Chen

Model	Equation	R^2(%)
Monod	$U^{-1} = 2025\ S^{-1} + 3.61$	99.6
	$K_s = 498$	
	$K = 0.350$	
	$\mu_{Max} = 0.284$	
Contois	$U^{-1} = 0.306\ X\ S^{-1} + 2.78$	99.1
	$B = 0.111$	
	$u_{Max} = 0.344$	
	$a = 0.115$	
	$\mu_{Max} = 0.377$	
	$K = 0.519$	
Chen & Hashimoto	$U^{-1} = 0.0190\ S_o\ S^{-1} + 3.77$	99.5
	$K = 0.006$	
	$a = 0.006$	
	$\mu_{Max} = 0.291$	
	$K = 0.374$	

Table 3. Summary of the three known substrate utilization models application.

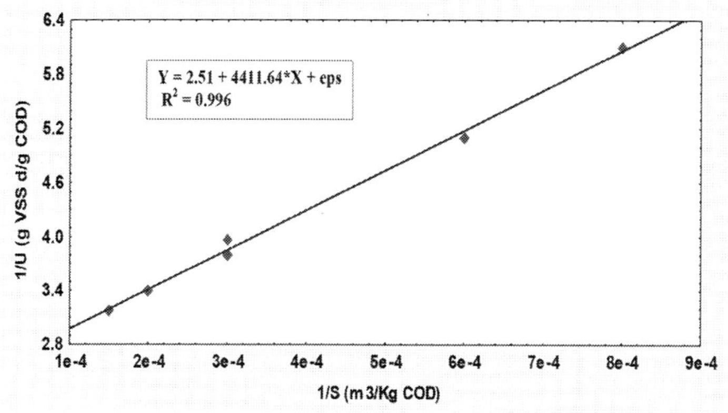

Figure 6. The Monod model.

and Hashimoto models performed better, shown that UMAS reactor performance should consider organic loading rates. These two models suggested that the predicted permeate COD concentration (S) is a function of influent COD concentration (S_o).

The percentage removal of COD by UMAS at various HRTs was shown in **Figure 9**. It was observed that COD removal efficiency increased as HRT increased from 8.6 to 500.8 days and it was in the range of 92.8–98.3%. It was found that this value higher than the 85% COD removal is observed for POME wastewater treatment using anaerobic fluidized bed reactors [34] and the 91.7–94.2% removal is observed for palm oil mill effluent wastewater treatment using membrane anaerobic system [35], and the 93.6–97.5% removal is observed for POME treatment using membrane anaerobic system [33]. Interestingly, it was found that there is no much difference in COD removal efficiency between HRTs of 500.8 days (98.3%) and 14.7 days

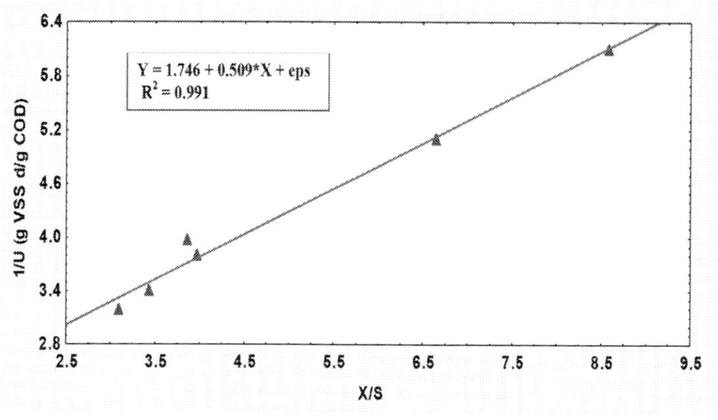

Figure 7. The Contois model.

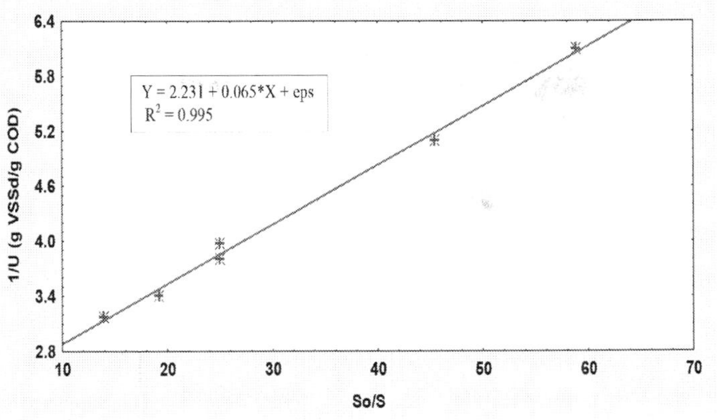

Figure 8. The Chen and Hashimoto model.

(96.0%). On the other hand, the COD removal efficiency has declined at shorter hydraulic retention time; at HRT of 8.6 days, the COD removal efficiency was reduced to 92.8%. **Table 2** results show that UMAS result might because of grown of volatile fatty acids inside the reactor. Usually, the hydraulic retention times were mainly effected by the ultrafiltration (UF) membrane influx rates, which directly determined the volume of influent (POME) that can be fed to the reactor.

3.2. Evaluation of UMAS biokinetic coefficients

The evaluated biokinetic coefficients based on COD basis by UMAS were analyzed as shown in **Table 2**.

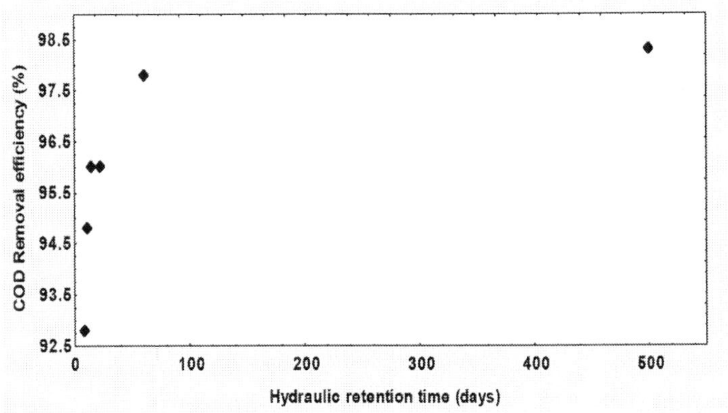

Figure 9. COD removal efficiency of UMAS under steady-state conditions with various hydraulic retention times.

The kinetic coefficients were calculated and summarized in **Table 3**. The growth yield coefficient, Y, value ranges from 0.32 to 0.68 gm VSS/gm COD, specific microorganic decay rate, b, and maximum substrate utilization rate, K, ranges from 0.350 to 0.374 COD/g VSS.day. **Figure 10** depicts the relationship between the substrate utilization rates (SUR) and the specific substrate utilization rate for COD with various hydraulic retention times. The HRTs range from 8.6 to 500.8 days. The biokinetic coefficients of growth yield, Y, and specific microorganic decay rate, b, were calculated from the slope and intercept as shown in **Figures 11** and **12**. The evaluated maximum specific biomass growth rates, $\mu_{max,}$ range from 0.248 to 0.474 day^{-1}.

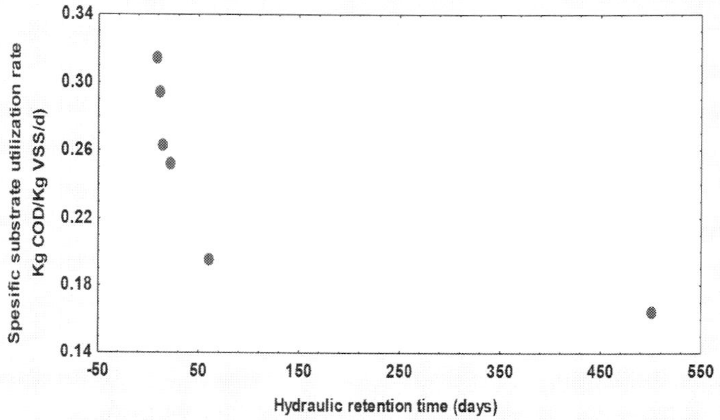

Figure 10. The specific substrate utilization rate for COD with various hydraulic retention times.

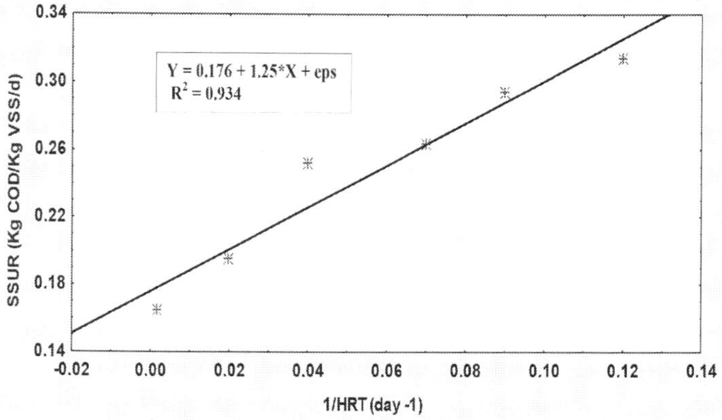

Figure 11. Evaluation of the growth yield, Y, and the specific biomass decay rate, b.

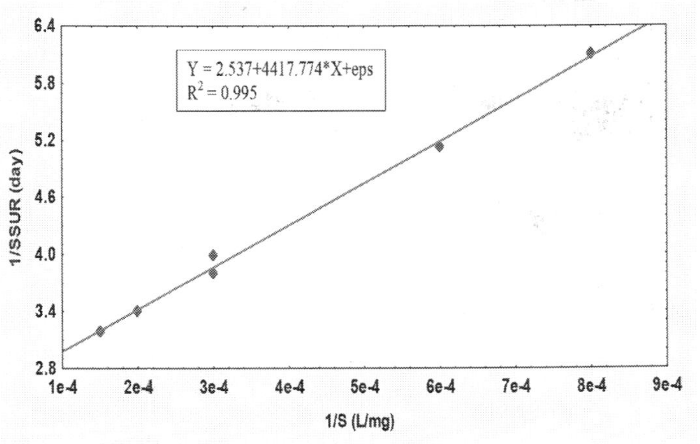

Figure 12. Evaluation of the maximum specific substrate utilization and the saturation constant, K.

4. Production of methane (CH_4) and carbon dioxide (CO_2) gases

A semicontinuous operation was conducted to verify the performance of the integrated ultrasonic membrane anaerobic system (UMAS) throughout a different hydraulic retention times (HRTs) and influent COD concentrations. In this study, the influent COD concentration was increased from 70,400 to 90,200 mg/l (for the six steady states). **Figure 13** illustrates the gas production rate and the methane content of the biogas. It was clear that the methane CH_4 yield decreased with increasing OLRs. Methane gas contents were varied from 64.6 to 81%, and the methane yield was varied from 0.39 to 0.70 CH_4/g COD/day. The decreased CH_4 yield with increasing OLR was also

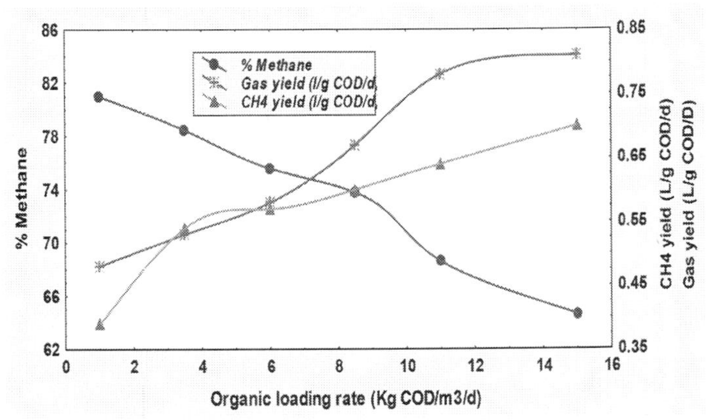

Figure 13. Gas production and methane content.

noted in many previous studies [36–40]. One of the reasons might be that shorter HRT of the system contributed to more active methanogens that were washed out during the removal of effluent. The gas production has increased from 290 to 540 L per day during the study. Biogas production increased with increasing OLRs from 0.48 l/g COD/day at 1.0 kg COD/m^3/day to 0.81 l/g COD/day at 15 kg COD/m^3/day. These findings are in line with the results obtained from Refs. [41–43].

5. Conclusions

The kinetic performance of newly designed ultrasonic membrane anaerobic system (UMAS) was evaluated in the treatment of palm oil mill effluent (POME).

The steady-state performance of ultrasonic membrane anaerobic system (UMAS) was evaluated under different influent COD concentrations (70,400–90,200 mg/l), hydraulic retention times (HRTs) (500.8–14.7 days) and OLR of 1.5–9.0 kg COD/m^3/day.

Among the three models applied, the Monod and Chen and Hashimoto models performed better, shown that UMAS reactor performance should consider organic loading rates. These two models suggested that the predicted permeate COD concentration (S) is a function of influent COD concentration (S_o).

It was observed that COD removal efficiency increased as HRT increased from 8.6 to 500.8 days, and it was in the range of 92.8–98.3%. The evaluated maximum specific biomass growth rates, μ_{max}, range from 0.248 to 0.474 day^{-1}.

It was found that the methane CH_4 yield decreased with increasing OLRs. Methane gas contents were varied from 64.6 to 81%, and the methane yield was varied from 0.39 to 0.70 CH_4/g COD/day.

Author details

Abdurahman Hamid Nour[1]* and Azhari Hamid Nour[2]

*Address all correspondence to: nour2000_99@yahoo.com

1 Faculty of Chemical and Natural Resources Engineering, University Malaysia Pahang-UMP, Pekan, Pahang, Malaysia

2 Faculty of Pure and Applied Sciences International University of Africa, Khartoum, Sudan

References

[1] Jundika, C. K., Sachin, V. J., Saad, A., Agus, P. S., Arun, S. M. (2016). Advances in biofuel production from oil palm and palm oil processing wastes: a review. *Biofuel Research Journal, 9*, 332–346.

[2] Hansen, S. B., Padfield, R., Syayuti, K., Evers, S., Zakariah, Z., Mastura, S. (2015). Trends in global palm oil sustainability research. *Journal of Cleaner Production, 100,* 140–149.

[3] MPOB, Malaysian Palm Oill Board. (2015). Applications. Retrieved from http://www.palmoilworld.org/applications.html.

[4] USDA, US Department of Agriculture. (2016). Global Consumption of Vegetable Oils from 1995/1996 to 2014/2015, by Oil Type (in Million Metric Tons). *PNG image (.png)* (Vol. 55.5 KB (56,904 bytes), pp. Vegetable Oils: Global Consumption by Oil Type 1995–2015). United States: Statista, USDA, Foreign Agricultural Service, Oilseeds: World Markets and Trade.

[5] Meryana, E., Budiono, A., Sutyanto, D. (2016). Palm Oil. Retrieved from http://www.indonesia-investments.com/business/commodities/palm-oil/item166.

[6] Sayer, J., Ghazoul, J., Nelson, P., Boedhihartono, A. K. (2012). Oil palm expansion transforms tropical landscapes and livelihoods. *Global Food Security, 1,* 114–119.

[7] Mba, O. I., Dumontn, M. J., Ngadinn, M. (2015). Palm oil: processing, characterization and utilization in the food industry—a review. *Food Bioscience, 10,* 26–41.

[8] Abdullah,A. L., Idris, A., Ahmadun, F. R., Baharin, B. S., Emby, F., Noor,M. M., Nour, A. H. (2005). A kinetic study of a membrane anaerobic reactor (MAR) for treatment of sewage sludge. *Desalination, 183,* 439–445.

[9] Yusoff, S. (2006). Renewable energy from palm oil—innovation on effective utilization of waste. *Journal of Cleaner Production, 14,* 87–93.

[10] Barrientos, M., Soria, C. (2016). Palm oil production by country in 1000 MT. Retrieved from http://www.indexmundi.com/agriculture/?commodity=palm-oil&graph=production

[11] Basiron, Y. (2007). Palm oil production through sustainable plantations. *European Journal of Lipid Science and Technology, 109,* 289–295.

[12] Carter, C., Finley, W., Fry, J., Jackson, D., Willis, L. (2007). Palm oil markets and future supply. *European Journal of Lipid Science and Technology, 109,* 307–314.

[13] Ma, A. N. (2000). Environment management for the palm oil industry. *Palm Oil Development, 30,* 1–10.

[14] Chin, K.K., Wong, K.K. (1982). Performance of an aerobic treatment system for palm oil sludge. In: M.Moo-Young (Ed.), Proceeding on Waste Treatment and Utilization-Theory and Practice of Waste Management. Waterloo, Canada, 56-64.

[15] Chin, K. K., Lee, S. W., Mohammad, H. H. (1996). A study of palm oil mill effluent treatment using a pond system, *Water Science and Technology, 34,* 119–123.

[16] Chan, K. S., Choi, C. F. (1982). Ponding System for Palm Oil Mill Effluent Treatment PORIM Regional Workshop on Palm Oil Mill Technology and Effluent Treatment, Kuala Lumpur, Malaysia.

[17] Ismail, S. (2005). Membrane separation technology for palm oil mill effluent (POME) treatment: an integrated approach. Ph.D. thesis, Universiti Sains Malaysia. 149-206.

[18] Thani, M. I., Hussin, R., Ibrahim, W. W. R., Sulaiman, M. S. (1999). Industrial Processes and the Environment: Crude Palm Oil Industry. Handbook No. 3, Department of Environment, Kuala Lumpur, pp. 7–54.

[19] Najafpour, G. D., Zinatizadeh, A. A. L., Mohamed, A. R., Isa, M. H., Nasrollahzadeh, H. (2006). High-rate anaerobic digestion of palm oil mill effluent in an up-flow sludgefixed film bioreactor. *Process Biochemistry, 41*, 370–379.

[20] Mahabot, S., Harun, A. R. (1986). Effluent treatment system for FELDA's palm oil mill. Technical Service Department, Kuala Lumpur.

[21] Oswal, N., Sarma, P. M., Zinjarde, S. S, Pant, A. (2002). Palm oil mill effluent treatment by a tropical marine yeast. *Bioresource Technology, 85*, 35–37.

[22] Monod, J. (1949). Growth of bacteria cultures. *Annual Review of Microbiology, 3*, 371–394.

[23] Contois, D. E. (1959). Kinetics of bacteria growth: relationship between population density and space growth rate of continuous cultures. *The Journal of General and Applied Microbiology, 21*, 40–50.

[24] Chen, Y. R., Hashimoto, A. G. (1980). Substrate utilization kinetic model for biological treatment processes. Biotechnology and Bioengineering, *22*, 2081–2095.

[25] Rahayu, A. S., Karsiwulan, D., Yuwono, H., Trisnawati, I., Mulyasari, S., Rahardjo, S., Hokermin, S., Paramita, V. (2015). Handbook POME-to-Biogas Project Development in Indonesia. In: B. Castermans, et al. Eds. 2nd ed., Winrock International, United States of America. 8-19.

[26] Youngsukkasem, S., Barghi, H., Rakshit, S. K., Taherzadeh, M. J. (2013). Rapid biogas production by compact multi-layer membrane bioreactor, efficiency of synthetic polymeric membranes. *Energies, 6*, 6211–6224.

[27] Yunus, A., Zahira, Y., Parul, A., Kamaruzzaman, S. (2015). Production of biogas and performance evaluation of existing treatment processes in palm oil mill effluent (POME). *Renewable and Sustainable Energy Reviews, 42*, 1260–1278.

[28] Gerardi, M. H. (2003). The Microbiology of Anaerobic Digester. Wiley-Interscience, New Jersey, pp. 51–57.

[29] Mes, T. Z. D., Stams, A. J. M., Reith, J. H., Zeeman, G. (2003). Methane production by anaerobic digestion of wastewater and solid wastes. In: J. H. Reith, et al. (Eds.), *Biomethane and Biohydrogen: Status and Perspectives of Biological Methane and Hydrogen Production*. Sustainable, Sanitation and Water Management, Switzerland, pp. 58–94.

[30] Girard, M., Palacios, J. H., Belzile, M., Godbout, S., Pelletier, F. (2013). Biodegradation in Animal Manure Management. In: Dr. R. Chamy (Ed.) *Biodegradation – Engineering and Technology* Retrieved from http://www.intechopen.com/books/biodegradation-engineering-and-technology/biodegradation-in-animal-manure-management. doi:10.5772/56151

[31] APHA. (2005). Standard Methods for the Examination of Water and Wastewater. American Public Health Association, New York.

[32] Nagano, A., Arikawa, E., Kobayashi, H. (1992). The treatment of liquor wastewater containing high-strength suspended solids by membrane bioreactor system. *Water Science and Technology, 26*(3–4), 887–895.

[33] Abdurahman, H. N., Yunus, R. M., Azhar H. N. (2011). Development of a membrane anaerobic system (MAS) for palm oil mill effluent (POME) effluent. *Desalination, 266,* 208–212.

[34] Idris, A. B., Al-Mamun, A. (1998). Effect of scale on the performance of anaerobic fluidized bed reactors (AFBR) treating palm oil mill effluent. In: Proceedings of Fourth International Symposium on Waste Management Problems in Agro-Industry, Istanbul, Turkey, pp. 206–211.

[35] Fakhru'l-Razi, A. (1994). Ultrafiltration membrane separation for anaerobic wastewater treatment. *Water Science and Technology, 30*(12), 321–327.

[36] Kim, H. W., Han, S. K., Shin, H. S. (2006). Simultaneous treatment of sewage sludge and food waste by the unified high-rate anaerobic digestion system. *Water Science and Technology, 53*(6), 29–35.

[37] Linke, B. (2006). Kinetic study of thermophilic anaerobic digestion of solid wastes from potato processing. *Biomass Bioenergy, 30,* 892–896.

[38] Ahmad, A. L., Ismail, S., Ibrahim, N., Bhatia, S. (2003). Removal of suspended solid and residual oil from palm oil mill effluent. *Journal of Chemical Technology and Biotechnology, 78*(9), 971–978.

[39] Ahmad, A. L., Ismail, S., Bhatia, S. (2005). Optimization of coagulation-flocculation process for palm oil mill effluent using response surface methodology. *Journal of Environmental Science and Technology, 39*(8), 2828–2834.

[40] Ahmad, A. L., Ismail, S., Bhatia, S. (2005). Ultrafiltration behavior in the treatment of agroindustry effluent: Pilot Scale Studie. *Chemical Engineering Science, 60*(19), 5385–5394.

[41] Awalludin, M. F., Sulaiman, O., Hashim, R., Nadhari, W. N. A. W. (2015). An overview of the oil palm industry in Malaysia and its waste utilization through thermochemical conversion, specifically via liquefaction. *Renewable and Sustainable Energy Reviews, 50,* 1469–1484.

[42] Grady, C. P. L., Lim, H. C., (1980) Biological Wastewater Treatment: Theory and Applications. Macel Dekker Inc., New York, pp. 220–222, 870–876.

[43] Zinatizadeh, A. A. L., Mohamed, A. R., Najafpour, G. D. (2006). Kinetic evaluation of palm oil mill effluent digestion in a high rate up-flow anaerobic sludge fixed film bioreactor. *Process Biochemistry Journal,* 41, 1038–1046.